ENERGY SCIENCE, ENGINEERING AND TECHNOLOGY

ENERGY AND WATER SECTOR INTERDEPENDENCE

A PRIMER

ENERGY SCIENCE, ENGINEERING AND TECHNOLOGY

Additional books in this series can be found on Nova's website under the Series tab.

Additional e-books in this series can be found on Nova's website under the e-book tab.

ENERGY SCIENCE, ENGINEERING AND TECHNOLOGY

ENERGY AND WATER SECTOR INTERDEPENDENCE

A PRIMER

WILFRED D. MEADOW
EDITOR

New York

For permission to use material from this book please contact us:
Telephone 631-231-7269; Fax 631-231-8175
Web Site: http://www.novapublishers.com

Library of Congress Cataloging-in-Publication Data

ISBN: 978-1-63117-789-7

Published by Nova Science Publishers, Inc. † New York

CONTENTS

PREFACE

This book provides background on energy for facilities that treat and deliver water to end users and also dispose of and discharge wastewater.

Chapter 1 – Water and energy are resources that are reciprocally and mutually linked, because meeting energy needs requires water, often in large quantities, for mining, fuel production, hydropower, and power plant cooling, and energy is needed for pumping, treatment, and distribution of water and for collection, treatment, and discharge of wastewater. This interrelationship is often referred to as the energy-water nexus, or the water-energy nexus. There is growing recognition that "saving water saves energy." Energy efficiency initiatives offer opportunities for delivering significant water savings, and likewise, water efficiency initiatives offer opportunities for delivering significant energy savings. In addition, saving water also reduces carbon emissions by saving energy otherwise generated to move and treat water.

This report provides background on energy for facilities that treat and deliver water to end users and also dispose of and discharge wastewater. Energy use for water is a function of many variables, including water source (surface water pumping typically requires less energy than groundwater pumping), treatment (high ambient quality raw water requires less treatment than brackish or seawater), intended end-use, distribution (water pumped long distances requires more energy), amount of water loss in the system through leakage and evaporation, and level of wastewater treatment (stringency of water quality regulations to meet discharge standards). Likewise, the intensity of energy use of water, which is the relative amount of energy needed for a task such as pumping water, varies depending on characteristics such as topography (affecting groundwater recharge), climate, seasonal temperature, and rainfall. Most of the energy used for water-related purposes is in the form of electricity. Estimates of water-related energy use range from 4% to perhaps 13% of the nation's electricity generation, but regional differences can be significant. In California, for example, as much as 19% of the state's electricity consumption is for pumping, treating, collecting and discharging water and wastewater.

Chapter 2 – Water and energy are critical resources that are reciprocally linked; this interdependence is often described as the water-energy nexus. Meeting energy-sector water needs, which are often large, depends upon the local availability of water for fuel production, hydropower generation, and thermoelectric power plant cooling. The U.S. energy sector's use of water is significant in terms of water withdrawals and water consumption. Thermoelectric cooling represented 41% of water withdrawn nationally, and the broader energy sector's water use (including biofuels) represented around 14% of water consumed nationally. Energy-related water consumption is anticipated to continue to increase in coming decades as the result of more domestic biofuel and unconventional onshore oil and natural gas production. Policy makers at the federal, state, and local levels are faced with deciding whether to respond to the growing water needs of the energy sector, and if so, which policy levers to use (e.g., tax incentives, loan guarantees, permits, regulations, planning, or education). Many U.S. energy sector water decisions are made by private entities, and state entities have the majority of the authority over water use and allocation policies and decisions.

For fuel production, water is either an essential input or is difficult and costly to substitute, and degraded water is often a waste byproduct that creates management and disposal challenges. U.S. unconventional oil and natural gas production has expanded quickly since 2008, and U.S. natural gas and coal exports may rise. This has sparked interest in the quantities of water and other inputs "embedded" in these resources, as well as the wastes produced (e.g., wastewaters from oil and natural gas extraction) and how they are reused or disposed (e.g., concerns over induced seismicity from disposal of oil and natural gas wastewaters). Much of the growth in water demand for unconventional fuel production is concentrated in regions with already intense competition over water (e.g., tight gas and other unconventional production in Colorado, Eagle Ford shale gas and oil in south Texas), preexisting water concerns (e.g., groundwater decline in North Dakota before Bakken oil development), or regions with abundant, but ecologically sensitive surface water resources (e.g., Marcellus shale region in Pennsylvania and New York).

Chapter 3 – The energy choices before Congress represent vastly different demands on domestic freshwater because water is used in varying amounts in most aspects of the energy sector. Transitions in the energy sector, such as the pursuit of greater energy independence and security, produce changes in how much and where the energy sector uses water. The energy sector is the fastest-growing water consumer in the United States, in part because of energy policies. Whether the federal government addresses the energy sector's rising water demand, and if so how, is one of the many energy decisions that may be considered by the 112th Congress.

Much of the growth in the energy sector's water demand is concentrated in regions with already intense competition over water. Whether the energy sector exacerbates or alleviates future water tensions is influenced by near-term energy policy and investment decisions. These decisions also may determine whether water will limit or harm U.S. capability to reliably meet the nation's energy demand. Part of the policy issue for Congress is identifying the extent of the federal role in responding to the energy sector's water use. Currently, the energy industry and states have the most responsibility for managing and meeting the energy sector's water demand.

In: Energy and Water Sector Interdependence
Editor: Wilfred D. Meadow
ISBN: 978-1-63117-789-7

Chapter 1

ENERGY-WATER NEXUS: THE WATER SECTOR'S ENERGY USE*

Claudia Copeland

SUMMARY

Water and energy are resources that are reciprocally and mutually linked, because meeting energy needs requires water, often in large quantities, for mining, fuel production, hydropower, and power plant cooling, and energy is needed for pumping, treatment, and distribution of water and for collection, treatment, and discharge of wastewater. This interrelationship is often referred to as the energy-water nexus, or the water-energy nexus. There is growing recognition that "saving water saves energy." Energy efficiency initiatives offer opportunities for delivering significant water savings, and likewise, water efficiency initiatives offer opportunities for delivering significant energy savings. In addition, saving water also reduces carbon emissions by saving energy otherwise generated to move and treat water.

This report provides background on energy for facilities that treat and deliver water to end users and also dispose of and discharge wastewater. Energy use for water is a function of many variables, including water source (surface water pumping typically requires less energy than groundwater pumping), treatment (high ambient quality raw

* This is an edited, reformatted and augmented version of a Congressional Research Service publication, CRS Report for Congress R43200, prepared for Members and Committees of Congress, from www.crs.gov, dated January 3, 2014.

water requires less treatment than brackish or seawater), intended end-use, distribution (water pumped long distances requires more energy), amount of water loss in the system through leakage and evaporation, and level of wastewater treatment (stringency of water quality regulations to meet discharge standards). Likewise, the intensity of energy use of water, which is the relative amount of energy needed for a task such as pumping water, varies depending on characteristics such as topography (affecting groundwater recharge), climate, seasonal temperature, and rainfall. Most of the energy used for water-related purposes is in the form of electricity. Estimates of water-related energy use range from 4% to perhaps 13% of the nation's electricity generation, but regional differences can be significant. In California, for example, as much as 19% of the state's electricity consumption is for pumping, treating, collecting and discharging water and wastewater.

Energy consumption by public drinking water and wastewater utilities, which are primarily owned and operated by local governments, can represent 30-40% of a municipality's energy bill. At drinking water plants, the largest energy use (about 80%) is to operate motors for pumping. At wastewater treatment plants, aeration, pumping, and solids processing account for most of the electricity that is used. Energy is the second highest budget item for these utilities, after labor costs, so energy conservation and efficiency are issues of increasing importance to many of them. Opportunities for efficiency exist in several categories, such as upgrading to more efficient equipment, improving energy management, and generating energy on-site to offset purchased electricity. However, barriers to improved energy efficiency by water and wastewater utilities exist, including capital costs and reluctance by utility officials to change practices or implement new technologies.

Topics for research to better understand water-related energy use include studies of energy demands for water at local, regional, and national scales; development of consistent data collection methodology to track water and energy data across all sectors; development and implementation of advanced technologies that save energy and water; and analysis of incentives, disincentives, and lack of incentives to investing in cost-effective energy or water efficiency measures.

INTRODUCTION

Water and energy are critical resources that are reciprocally and mutually linked. Meeting energy needs depends upon the availability of water, often in large quantities, for mineral extraction and mining, fuel production, hydropower, and thermoelectric power plant cooling.[1] Likewise, energy is required for the pumping, conveyance, treatment and conditioning, and distribution of water and for collection, treatment, and discharge of wastewater. This interdependence, which is often described as the water-energy nexus, or energy-water nexus, is illustrated in the following graphic from a U.S. Department of Energy report.

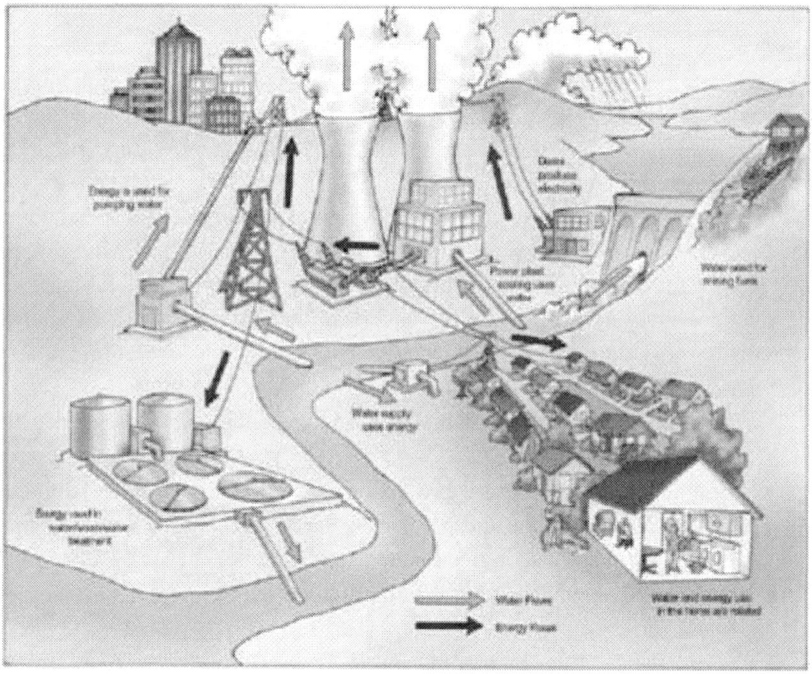

Source: U.S. Department of Energy, Energy Demands on Water Resources, Report to Congress on the Interdependency of Energy and Water, December 2006, p. 13.

Figure 1. Examples of Interrelationships Between Water and Energy.

This report first discusses water-related energy use broadly and then energy for facilities that treat and deliver water to end users and also dispose of and discharge wastewater. There is growing recognition that "saving energy saves water," and the report describes options and impediments for energy efficiency by these facilities. It also identifies several areas of research and information needs concerning energy for water uses.

Energy for Water Use

In the United States, more than 400 billion gallons of water is withdrawn daily from surface and ground water sources of freshwater and saline-water to supply domestic uses, agriculture including irrigation, industry, mining, and thermoelectric power.[2] Information about the energy— especially electricity— that is needed to pump, transport, deliver, and process that water is fragmentary and not well documented overall. In particular, as described further below, energy needs for self-supplied domestic, industrial, and energy water is largely unknown, but are likely to be large. Interest has been growing in better understanding of the energy-related needs of providing water to diverse sectors of the economy.

In a 2002 report, the Electric Power Research Institute (EPRI) estimated that 4% of the nation's electricity use goes towards moving and treating water and wastewater by public and private entities.[3] Today that estimate is considered a good starting place for understanding the magnitude of energy demands for providing these water services, but deficient in several respects.

- It relied on secondary source data that were then well over a decade old.
- It did not include future projections of electricity requirements for water supplies in the thermoelectric sector (because it assumed that energy for water use in this sector would decline).
- It did not consider on-site heating, cooling, pumping and softening of water for end-use.
- It did not consider that in the future a large proportion of new water demands will be met by sources with greater energy intensities, such as groundwater pumped from greater depths and seawater desalination.

More recently, others have attempted to expand on the EPRI analysis, using additional and updated data from a variety of sources, to develop a baseline estimate of water-related energy use in the United States. These analyses suggest that energy for publicly supplied water and wastewater is a larger share of U.S. energy use than EPRI estimated. For example, a 2009 report by the River Network, a national advocacy group for freshwater conservation and watershed restoration, estimated that water-related energy use, including for heating in the residential and commercial sectors, was 52.1 billion killowatt hours (kWh), equivalent to 13% of U.S. electricity consumption in 2007.[4] National data can obscure differences in water-related energy use that are regional or state-specific, as reflected in a 2005 study by the California Energy Commission, which found that "water-related energy use [in California] consumes 19 percent of the state's electricity, 30 percent of its natural gas, and 88 billion gallons of diesel fuel every year – and this demand is growing." Pumps that move water from the San Joaquin Valley to southern California for domestic and irrigation water uses are the single largest power load in the state.[5]

Researchers at the University of Texas at Austin have attempted to quantify the energy embedded in the U.S. public water supply, which is the primary water source of residential, commercial, and municipal users. One such analysis concluded that energy use associated with the public water supply is 4.1% of the nation's annual primary energy consumption and 6.1% of national electricity consumption, but this analysis excluded energy requirements associated with water for agriculture, industrial, and self-supplied sectors (e.g., agriculture, thermoelectric, and mining). In this analysis, electricity consumption by public drinking water and wastewater utilities for pumping, conveyance, treatment, distribution, and discharge was 56.6 billion kWh, or 11.5% of primary energy and 21.6% of electricity consumption for water end-use, respectively, in 2009.[6]

A second analysis by these researchers looked more broadly at energy needs for water supply, adding industrial and thermoelectric sectors to others considered previously.[7] Water-related energy use throughout the economy varies across sectors, and analysis of some sectors is complex and limited by incomplete data (e.g., cooking-related activities vary across residences and are not well documented). This analysis concluded that direct water-related energy consumption was 12.6% of national primary energy consumption in 2010. This amount of energy, 12.3 quadrillion BTUs, is the equivalent of annual energy consumption of about 40 million Americans. It also estimated that energy losses at the point of electricity generation, transmission and distribution, and end-use represent 58% of the total primary energy that was consumed for water-related purposes, reflecting varying efficiencies of water heating and boiler technologies. The estimate of waste heat losses is subject to uncertainty, the researchers said.

Because of inadequate data and other factors, missing from some analyses is water-related energy for several important end-use sectors.

- Self-supplied water, which is a high percentage of power plant use and of some industrial uses such as mining. Some of this energy-for-water is in the form of electricity, but much of it is likely direct use of fuels on-site. Privately operated residential water supply wells also utilize energy for pumping, none of which is accounted for.

- Agricultural use of water for livestock and irrigation—second in volume only to water use for thermoelectric power, according to the U.S. Geological Survey—is generally omitted from these analyses, although substantial energy, which generally is self-supplied, is needed for pumping.
- The transportation sector, although the majority of energy consumed is for petroleum-based transportation fuels, which is presumably reflected in water-forenergy analyses.
- The bottled water industry, which is a substantial drinking water source in the United States and consumes energy to collect, treat, bottle, and distribute its products.

Energy use for water is a function of many variables, including water source (surface water pumping typically requires less energy than groundwater pumping), treatment (high ambient quality raw water requires less treatment than brackish or seawater), intended end-use, distribution (water pumped long distances requires more energy), amount of water loss in the system through leakage and evaporation, and level of wastewater treatment (stringency of water quality regulations to meet discharge standards). Likewise, the intensity of energy use of water[8] varies depending on characteristics such as topography (affecting groundwater recharge), climate, seasonal temperature, and rainfall. Focusing on national data can mask large variations, because the United States is a difficult country to generalize.

For example, the lifecycle energy-intensity of water in cities nationally is estimated to be 3,300-3,600 kWh per million gallons delivered and treated, but ranges from 2,700 kWh/million gallons in New York, New York, to 5,000 kWh per million gallons in Austin, Texas.[9] Energy intensity varies within states, as well. In California, the energy intensity of the water use cycle ranges from 4,000 kWh per million gallons in the northern part of the state to 12,700 kWh per million gallons in southern California, reflecting differences in the volume of water pumped, lifted, and transported hundreds of miles and over mountains from points of collection to points of need in the southern part of the state.[10] The energy intensity of a particular activity's water use, also described as the embedded energy of the activity, can have disproportionate impacts elsewhere.

For example, policies that promote the use of energy-intensive water supply such as pumping and distributing water over long distances, rather than policies that promote water conservation, water reuse, or aquifer recharge, adversely impact one sector to serve another.[11]

In every sector, there are opportunities for practices that would save energy and also save water. The Environmental Protection Agency's (EPA) WaterSense program promotes this concept by emphasizing that "saving water saves energy."[12]

Energy efficiency initiatives offer opportunities for delivering significant water savings, and likewise, water efficiency initiatives offer opportunities for delivering significant energy savings. In the commercial, industrial, and institutional sectors, potential water savings through energy efficiency and other measures could be 15-30% without reducing the services derived from the water. The potential for significant water and energy savings also exists in other sectors such as agriculture.[13]

Generating the energy associated with water use also produces carbon dioxide (CO_2) emissions that contribute to climate change. It has been estimated that water-related carbon emissions in 2005 were approximately 290 million metric tons, or 5% of all U.S. carbon emissions. Water-related CO_2 emissions were equivalent to the annual greenhouse gas emissions of 53 million passenger vehicles. By sector, water heating was responsible for 70% of the water-related carbon emissions, wastewater treatment was responsible for 18%, water supply was responsible for 8%, and agricultural activities were responsible for 6%.[14] Thus, saving water saves energy and also reduces carbon emissions.[15]

Energy for Water Supply and Wastewater Facilities

There are about 200,000 drinking water treatment systems in the country, of which about 52,000 are community water systems that serve 25 or more year-round residents.

Most U.S. drinking water is provided by relatively large community water systems. Nearly 85% of the U.S. population is supplied by about 5% of these systems; the remaining 95% include a large number of small and very small systems serving 3,300 persons or fewer.

Public agencies own and operate most community water systems; a small number are privately operated. Smaller utilities use more electricity and pay more per unit of water produced than do medium and large utilities, due to economies of scale. Nearly all of the energy consumed is electricity, about 80% of which is used by motors for pumping.

There are approximately 15,000 U.S. wastewater treatment plants. Most are publicly owned, and they serve more than 75% of the U.S. population. Nearly 70% of facilities are small, serving only 10% of the U.S. population. Approximately 22% are large (with flow greater than 1 million gallons per day) and serve over 85% of the population. Wastewater systems generally consist of collection systems (sewers and pumping stations), treatment facilities, and effluent disposal. Like water supply utilities, nearly all of the energy consumed is electricity. Wastewater aeration, pumping, and solids processing account for most of the electricity used in wastewater treatment.

For both types of systems, greater amounts of energy are required for more advanced treatment levels. Similarly, the age of the system and equipment are important: as systems age, equipment decreases in efficiency, resulting in an increase in electricity requirements.

According to EPA, community drinking water and publicly owned wastewater systems use 75 billion kWh per year—as much as the pulp and paper and petroleum industries combined, or enough electricity to power 6.75 million homes.[16] Energy is the second highest budget item for municipal drinking water and wastewater facilities, after labor costs, with utilities spending about $4 billion a year. Energy consumption by drinking water and wastewater utilities can comprise 30-40% of a municipality's total energy bill.

Water utilities are highly regulated entities whose primary goals are to meet regulatory requirements for protecting public health and the environment and to provide services for reasonable and fair rates. The energy efficiency of these utilities generally has not been a primary goal or considered as an element of rate determinations. Nevertheless, as populations grow and environmental requirements become more stringent, demand for electricity at drinking water and wastewater utility plants is expected to grow by approximately 20%.

Moreover, as electricity rates increase, energy conservation and efficiency are issues of increasing importance to many utilities. By some estimates, potential energy savings by drinking water and wastewater utilities are in the range of 15-30% per year.[17] Opportunities for efficiency exist in several categories.

- *Optimizing system processes*, such as modifying pumping and aeration operations and implementing monitoring and control systems through SCADA (supervisory control and data acquisition) systems to increase the energy efficiency of equipment. EPRI has estimated that drinking water facilities can achieve energy savings of 5-15% through adjustable speed drives and high-efficiency motors and drives and 10-20% through process optimization and SCADA systems. In wastewater facilities, EPRI estimates that 10-20% energy savings are possible through process optimization.[18]

- *Upgrading to more efficient equipment and right-sizing equipment* for the capacity of the facility (plants and pipes often are oversized, to accommodate future peak load). Pumps and other equipment used beyond their expected life operate well below optimal efficiency. In addition, energy is embedded through pipe systems, since leaking drinking water pipes require more energy to deliver water to the end user. Leaky sewer lines allow groundwater to infiltrate and increase the flow of water into the wastewater treatment plant. All water systems have losses, which are cumulative along segments of the water-use cycle. Projects to address water loss and improve end-use efficiency can be promoted as both water- and energy-savings investments.
- *Improved energy management.* It is widely recognized that water utilities need to develop better understanding of their current energy use, and public and private research and programs have focused on this goal. For example, some states have developed programs to help water utilities better manage energy use. The New York State Energy Research and Development Authority has done extensive work to help water utilities benchmark their energy use and support a range of initiatives through its Focus on Municipal Water and Wastewater Treatment program. It developed a best practices handbook for the water and wastewater sectors, including methods to track performance and assess program effectiveness.[19] The California Energy Commission also has been active on energy management issues. It has reported on case studies of water facilities throughout the state that implemented energy efficiency measures, including resulting energy and cost savings. In 2006, the California Public Utilities Commission (CPUC) directed the state's largest electric utilities to partner with water agencies, undertake specific water conservation and efficiency programs, and measure the results.[20] Because one of the largest end uses of electricity in California is in treating, heating, and conveying water, energy companies in that state are working with the water sector to help utilities and companies reduce their energy use and

boost their efficiency. At the federal level, EPA has developed a number of tools for water utilities, including an energy management guidebook, energy conservation guides, and self-assessment and energy audit tools. Energy Star, a joint program of the Department of Energy and EPA, has developed a Portfolio Manager, an online benchmarking tool that allows drinking water and wastewater utilities to evaluate their energy use and compare their operations to similar facilities.[21] Others have contributed research on best practices, energy conservation, and benchmarking energy use, including the Water Environment Research Foundation, the Water Research Foundation, and the American Council for an Energy-Efficient Economy.

- Some water utilities are *generating energy on-site* to offset purchased electricity. Beyond efficiency measures, they illustrate ways in which water utilities are reducing their energy costs by recovering energy from municipal waste and using the resulting biogas to generate electricity, heat the plant, and in some cases sell electricity back to the grid. For example, D.C. Water, the wastewater utility in Washington, D.C., is constructing a project to convert residuals that remain after wastewater is treated into fuel for combined heat and power operations at the facility. The utility estimates that when the project is completed, it will save $10 million per year on electricity (powering one-third of the treatment plant) and another $10 million annually in solid waste management. Another example is the Gloversville-Johnstown, NY Joint Wastewater Treatment Facilities. In 2003, it began accepting dairy whey as a fuel for its own energy through a combined heat and power process. Reduced electrical costs save the utility about $500,000 per year, and accepting the dairy wastes results in additional revenue of $750,000 annually. Plants also are using other sources of renewable energy: solar panels at the Calera Creek Water Recycling Plant in Pacifica, California, provide 10-15% of the plant's energy needs and save an estimated $100,000 per year. Savings from such projects are contingent upon recouping capital investment costs.

Several barriers to improved energy efficiency by water and wastewater utilities are apparent. Many of them derive from the culture of water utilities and outside constraints placed on them.[22]

- Cost. Utilities have limited resources. Their capital and operations budgets are constrained, while the up-front costs of installing more energy-efficient equipment can be prohibitive. Some funding sources to finance projects do exist (e.g., private sector, municipal bonds), but may not be suitable or well known to water utility officials. Federal and state funding for energy efficiency projects is limited.

- Municipalities that own and operate water utilities generally are risk-averse, reluctant to change practices, and hesitant to implement new technologies. The tendency is to wait until equipment fails rather than be pro-active.
- Facility operators who could advocate for energy efficiency often are disconnected from those in the utility who pay the electricity bill. Most water and wastewater facilities were built decades ago when electricity costs were low enough to be of little concern. Facilities and equipment were designed to run continuously, without regard for wasted energy. To the extent that water utilities can pass on energy costs to customers, there may be little incentive to investigate energy efficiencies. Utility managers may not understand how energy is used at a plant and how to reduce, or even control, energy costs.

Other barriers stem from a lack of coherence and coordination between water and energy policies, planning, and decisionmaking roles. Energy and water decisions have historically been made independently of each other. Water planners typically assume that they have the energy that they need, and energy planners assume that they have the water that they need. Both are likely to use different strategic planning: private companies acting under market forces dictate the location of energy infrastructure, while water infrastructure is often located using public interest criteria. A mismatch in planning objectives by different actors can prevent the beneficial siting and combing of technologies. Likewise, water policy in the United States is usually structured in a bottom-up fashion with decisions driven by local water authorities, because water supply management is generally the responsibility of states. Energy policy, in contrast, is usually structured in a top-down fashion with federal agencies setting many standards and requirements.[23]

To overcome such barriers, a 2013 report recommends that the electricity, water supply, and wastewater sectors should foster cross-sector communication and engage in collaborative planning. To reduce potential disincentives and risks of cross-sector coordination, the report suggests that states could implement policies that incentivize wastewater plants to install energy projects and give credits to electric utilities for incorporating those generation sources into their portfolios.

Although some states do this now (for example, Massachusetts), the report states that new policies are needed to build the confidence of leaders in the water and electric power sectors and motivate them to "go beyond compliance."[24]

Research and Information Needs

Several areas of research and information needs concerning energy for water uses have been identified by a range of researchers and stakeholders.[25] They suggest:

- Data that could help decision makers and users fill what is now an incomplete picture of energy needs for water uses are lacking. This is apparent across sectors and also within individual sectors. The U.S. Geological Survey collects water use data, but national scale reports generally are issued only every five years,[26] the level of detail is limited, and related energy use is not considered. The Department of Energy collects energy data and forecasts energy use, but its work to address energy for water use is limited.[27] No public or private entity systematically collects energy data from public water utilities, users who self-supply their water, or other end users of water. Data that exist are scattered and often are not available at a scale needed by decision makers.
- More integrated research is needed on water and energy operations. Information is needed to understand where energy is used in water and wastewater infrastructure facilities, what opportunities for improvement exist, and how to establish priorities for action. Ideally, consistent data collection methodology is needed to gather and track water and energy data across all sectors and within sectors, such as at the utility level, and to aid benchmarking. Standards for data collection, coordination, and quality control are lacking.
- Research is needed on advanced technologies that save energy and save water, and partnerships between government and the private sector that move research and development from bench-scale to implementation are needed.
- Better understanding is needed of linkages between energy, water, land, and agriculture and risks of climate change and extreme weather events on water availability and energy supply.

- Policies and approaches are needed to encourage the water and energy sectors to move toward integrated resource management.
- Analysis is needed of incentives, disincentives, and lack of incentives to investing in cost-effective energy or water efficiency measures. One area of interest is regulatory barriers to co-implementation of efficiency programs in the water and energy sectors, such as policies in some states that restrict electric and gas utilities from utilizing biogas that is recovered from wastewater utilities.
- More education and outreach to all types of water users, the general public, and public officials are needed on the water-energy nexus and how improving efficiency involves the reciprocity of saving energy and saving water.

End Notes

[1] CRS Report R43199, Energy-Water Nexus: The Energy Sector's Water Use, discusses the related topic of water needs for the energy sector.

[2] Susan S. Hutson, Nancy L. Barber, and Joan F. Kenny, et al., Estimated Use of Water in the United States in 2000, U.S. Department of the Interior, U.S. Geological Survey, U.S. Geological Survey Circular 1268, 2004.

[3] Electric Power Research Institute, Water & Sustainability (Volume 4): U.S. Electricity Consumption for Water Supply & Treatment—the Next Half Century, 1006787, Topical Report, March 2002.

[4] Bevan Griffiths-Sattenspiel and Wendy Wilson, The Carbon Footprint of Water, River Network, May 2009. Hereinafter, The Carbon Footprint of Water.

[5] California Energy Commission, California's Water-Energy Relationship, Final Staff Report, CEC-700-2005-011-SF, November 2005, p. 1. Hereinafter, California's Water-Energy Relationship.

[6] Kelly M. Twomey and Michael E. Webber, "Evaluating the Energy Intensity of the US Public Water Supply," Proceedings of the ASME 2011 5th International Conference on Energy Sustainability, ES2011-54165, August 2011. Hereinafter, Twomey and Webber 2011.

[7] Kelly T. Sanders and Michael E. Webber, "Evaluating the energy consumed for water use in the United States," Environmental Research Letters, vol. 7, 034034 (2012).

[8] Energy intensity is the relative amount of energy needed to perform water management-related tasks such as treating and pumping water. It is typically expressed as the number of kilowatt-hours per million gallons of water.

[9] Twomey and Webber 2011, p. 8.

[10] California's Energy-Water Relationship, pp. 9-11.

[11] Carey W. King, Ashlynn S. Stillwell, and Kelly M. Twomey, et al., "Coherence Between Water and Energy Policies," Natural Resources Journal, vol. 53, no. 1 (Spring 2013), p. 185.Hereinafter, King, Stillwell, and Twomey 2013.

[12] The WaterSense program offers a label for products and services that are certified to save water without sacrificing performance, thus promoting water efficiency and supporting a

market for water-efficient products. WaterSense addresses residential and commercial water use. According to EPA, since the program's inception in 2006, WaterSense has helped consumers save a cumulative 487 billion gallons of water and $8.9 billion in water and energy bills. In addition, the Department of Energy has established minimum energy conservation standards for more than 50 categories of residential, commercial, and industrial appliances and equipment, including standards for several categories of appliances and devices that also specify maximum water use standards (showerheads and faucets; toilets and urinals; residential and commercial clothes washers; residential dishwashers; and prerinse spray valves). These standards were promulgated under the Energy Policy and Conservation Act of 1975, as amended.

[13] See Alliance for Water Efficiency, http://www.allianceforwaterefficiency.org/; the American Council for an Energy-Efficient Economy, http://www.acee.org; and The Carbon Footprint of Water, pp. 26-29.

[14] The Carbon Footprint of Water, p. 24. In this analysis, carbon emissions resulting from water supply and treatment assumed that all energy comes from electricity, with a carbon intensity of 1.36 lbs. CO2 /kWh. The carbon intensity of other energy sources varies, ranging from 0.12 lbs. CO2 per cubic foot of natural gas to 22.4 lbs. CO2 per gallon of fuel oil.

[15] See, for example, United Utilities, "Water and your carbon footprint," http://www.unitedutilities.com/water-andyour-carbon-footprint.aspx; and Wisconsin Department of Natural Resources, "Save Energy and Reduce Carbon Emissions," http://dnr.wi.gov/org/caer/ce/eek/teacher/globaltip.htm.

[16] Cheryl McGovern, U.S. Environmental Protection Agency, "Benchmarking Wastewater Facilities in Portfolio Manager," June 23, 2009, http://www.epa.gov/region9/water infrastructure/training/energy-workshop/docs/2009/ energystar-benchmark.pdf.

[17] Consortium for Energy Efficiency, CEE National Municipal Water and Wastewater Facility Initiative, January 1, 2010, p. 1.

[18] Ibid., p. 8.

[19] See http://www.nyserda.ny.gov/BusinessAreas/Energy-Efficiency-and-Renewable-Programs/Commercial-andIndustrial/Sectors/Municipal-Water-and-Wastewater-Facilities.aspx.

[20] ECONorthwest, Embedded Energy in Water Pilot Programs Impact Evaluation, Final Report, Study ID:CPU0053.01, March 9, 2011.

[21] See http://water.epa.gov/infrastructure/sustain/waterefficiency.cfm. The Portfolio Manager can be used by a range of facilities, including K-12 schools, hospitals, hotels, and offices, to track and assess energy and water consumption. It was extended to wastewater utilities in 2007 and later to drinking water utilities.

[22] See discussion in U.S. Government Accountability Office, Energy-Water Nexus, Amount of Energy Needed to Supply, Use, and Treat Water Is Location-specific and Can Be Reduced by Certain Technologies and Approaches, GAO-11-225, March 2011.

[23] King, Stillwell, and Twomey 2013, pp. 185-191.

[24] The Johnson Foundation at Wingspread, Building Resilient Utilities, How Water and Electric Utilities Can Co-Create their Futures, November 2013, p. 15.

[25] For discussion of research and information needs, see, for example, GEI Consultants, Water-Energy Nexus Research, Recommendations for Future Opportunities, Alliance for Water Efficiency and American Council for an Energy-Efficient Economy, Final Report, Project No. 130240, June 2013.

[26] The most recent national report was issued in 2004 and describes estimated U.S. use of water in 2000.

[27] DOE's Energy Information Administration (EIA) has conducted periodic Commercial Buildings Energy Consumption Surveys since the mid-1970s. A limited number of questions about water consumption were asked for the first time in the 2007 survey, including total volume of water consumed, cost, whether the volume was metered or estimated, and how much of the water was used outdoors. Sixteen commercial building categories (e.g., health care, lodging, food service) were surveyed. EIA reported on the results, without analysis, in 2012. See http://www.eia.gov/ consumption/commercial/reports /2007/hospital-water-data-collection.cfm.

In: Energy and Water Sector Interdependence
Editor: Wilfred D. Meadow
ISBN: 978-1-63117-789-7

Chapter 2

ENERGY-WATER NEXUS: THE ENERGY SECTOR'S WATER USE*

Nicole T. Carter

SUMMARY

Water and energy are critical resources that are reciprocally linked; this interdependence is often described as the water-energy nexus. Meeting energy-sector water needs, which are often large, depends upon the local availability of water for fuel production, hydropower generation, and thermoelectric power plant cooling. The U.S. energy sector's use of water is significant in terms of water withdrawals and water consumption. Thermoelectric cooling represented 41% of water withdrawn nationally, and the broader energy sector's water use (including biofuels) represented around 14% of water consumed nationally. Energy-related water consumption is anticipated to continue to increase in coming decades as the result of more domestic biofuel and unconventional onshore oil and natural gas production. Policy makers at the federal, state, and local levels are faced with deciding whether to respond to the growing water needs of the energy sector, and if so, which policy levers to use (e.g., tax incentives, loan guarantees, permits, regulations, planning, or education). Many U.S. energy sector water decisions are made by private entities, and

* This is an edited, reformatted and augmented version of a Congressional Research Service publication, CRS Report for Congress R43199, dated January 2, 2014.

state entities have the majority of the authority over water use and allocation policies and decisions.

For fuel production, water is either an essential input or is difficult and costly to substitute, and degraded water is often a waste byproduct that creates management and disposal challenges. U.S. unconventional oil and natural gas production has expanded quickly since 2008, and U.S. natural gas and coal exports may rise. This has sparked interest in the quantities of water and other inputs "embedded" in these resources, as well as the wastes produced (e.g., wastewaters from oil and natural gas extraction) and how they are reused or disposed (e.g., concerns over induced seismicity from disposal of oil and natural gas wastewaters). Much of the growth in water demand for unconventional fuel production is concentrated in regions with already intense competition over water (e.g., tight gas and other unconventional production in Colorado, Eagle Ford shale gas and oil in south Texas), preexisting water concerns (e.g., groundwater decline in North Dakota before Bakken oil development), or regions with abundant, but ecologically sensitive surface water resources (e.g., Marcellus shale region in Pennsylvania and New York).

Conventional hydropower accounts for approximately 8% of total U.S. net electricity generation, and more than 80% of U.S. electricity is generated at thermoelectric facilities that depend on cooling water. Water availability issues, such as regional drought, low flow, or intense competition for water, can curtail hydroelectric and thermoelectric generation. An assessment of the drought vulnerability of electricity in the western United States found broad resiliency, while also identifying the Pacific Northwest and the Texas grid at higher risk. Future withdrawals associated with electric generation may grow slightly, remain steady, or decline depending on a number of factors. These include reduced generation from facilities using once-through cooling because of compliance with proposed federal cooling water intake regulations or shifts in how electricity is generated (e.g., less from coal and more from wind and natural gas).

Energy choices represent complex tradeoffs; water use and wastewater byproducts are two of many factors to consider when making energy choices. For many policy makers, concerns other than water—low-cost reliable energy, energy independence and security, climate change mitigation, public health, and job creation—are more significant drivers of their positions on energy policies.

Water and energy are critical resources that are reciprocally linked. Energy is required for the pumping, conveyance, treatment and conditioning, and distribution of water and for collection, treatment, and discharge of wastewater. Likewise, as described in this report, meeting energy sector needs depends upon the local availability of water, often in large quantities, for mineral fuel production,[1] hydropower, and thermoelectric power plant cooling. This interdependence is often described as the water-energy nexus. This report addresses how the U.S. energy sector uses and relies on water; it provides summary descriptions divided into four topics: (1) Water for Energy Primer, (2) Fuel Production, (3) Electric Grid and Generation, (4) Policy Response Options and Considerations. CRS Report R43200, *Energy-Water Nexus: The Water Sector's Energy Use*, addresses the related topic of energy needs of the water sector.

WATER FOR ENERGY PRIMER

Energy-Sector Water Use and Vulnerability Is Receiving Increased Attention

Available projections estimate that, by 2030, U.S. water consumption will increase by 7% above the level consumed in 2005; 85% of this growth is attributed to the energy sector (including biofuels).[2] The U.S. energy sector's use of water is significant in terms of water withdrawals and water consumption.[3]

- **Energy Sector:** While agriculture dominates U.S. water consumption (71%), the energy sector (including biofuels, thermoelectric, and fuel production) is the second-largest consumer at 14%, and domestic and public uses are third at 7%.[4] Multiple factors contribute to the energy sector being the fastest-growing water consumer. Biofuels produced from irrigated feedstocks play a significant role, as well as expanding production of onshore unconventional oil and natural gas and hydro-stimulation of aging wells.
- **Electric Generation:** Water dependence is a risk for hydroelectric and thermoelectric generation. During low-flow or high-heat events, water intakes and high water temperatures may harm or limit thermoelectric cooling. In 2005, thermoelectric cooling water represented 41% of water withdrawn nationally,[5] and almost 6% of water consumed nationally.[6] Also, the withdrawal and discharge of cooling water can harm aquatic organisms.

- **Fuel Production:** Water is either an essential input or is difficult and costly to substitute; degraded water is often a waste byproduct. The potential for human- induced seismic events is receiving scientific and political attention because of concerns over the possible connection between oil and gas wastewater injection and the recent frequency of earthquakes, particularly in the central and eastern United States.
- **Efficiency and Conservation:** Reducing energy demand through energy and water efficiency[7] and more water-efficient generation (e.g., electricity from wind,[8] photovoltaics, or natural gas) can reduce water demand. Current water efficiency incentives in fuel production include minimizing water management costs and reducing operational disruptions.
- **Embedded Water:** U.S. unconventional oil and natural gas production has expanded quickly due to the combined use of hydraulic fracturing and horizontal drilling techniques for well development.[9] This expansion has sparked interest in the quantities of water and other inputs "embedded" in energy resources.

Relevant Data and Research Are Improving; Significant Gaps Remain

In 2012, the Government Accountability Office, in *Energy-Water Nexus: Coordinated Federal Approach Needed to Better Manage Energy and Water Tradeoffs*, stated that "making effective policy choices will continue to be challenging without more comprehensive data and research."[10] Improving data on water use by the energy sector is challenging for a number of reasons. For example, much of the U.S. energy sector is private; data consistency, accuracy, and currency are problematic; and it is costly to maintain high-quality data for an evolving and dispersed industry.

While data challenges exist, access to relevant research and data is improving. The Department of Energy (DOE) disseminates energy-water related studies on a public online platform.[11] The reports mentioned in the box below provide additional information on the energy-water nexus while also identifying areas needing improved understanding. While these reports differ in their focus, they each mention the stresses that climate change places on the energy-water nexus.

FUEL PRODUCTION

Unconventional Oil and Gas Production Often Concentrates Water Use Geographically and Temporally

Regional water resource opportunities and challenges for fuel production vary based on several factors, including (1) which fuel is being produced in the region, (2) the local and regional significance of its water use, and (3) regional conditions for management of wastewaters.

- Much of the growth in water demand for unconventional fuel production is concentrated in regions with already intense competition over water (e.g., tight gas and other unconventional production in Colorado, Eagle Ford shale gas and oil in south Texas), preexisting water concerns (e.g., groundwater decline in North Dakota before Bakken oil development), or abundant but ecologically sensitive surface water resources (e.g., Marcellus shale region in Pennsylvania and New York).
- The cumulative water needs of multiple drilling and fracturing operations may be locally or temporally significant. Often many shale gas, tight gas, and tight oil wells are located in close proximity to each other as a formation is developed, with many wells being drilled and fractured from the same location. Water use for these wells is concentrated in the early stages of well development, usually in the first few weeks. Once the well is producing, little or no water is required unless refracturing is performed. How much water is used for well development is highly variable both across and within formations.
- Data on source water remain sparse. Groundwater often is used for shale operations when it is available and access is permitted. Surface waters also are used, but may require transport by truck. In cases of limited water access, well developers also have obtained water by purchasing it from municipalities or paying individual land owners for their supplies.

Reports Linking Energy-Water Nexus to Climate Change

U.S. government reports and reports by stakeholders are increasingly addressing the links between energy, water, and climate change. Examples include the following reports:

- **January 2013, National Climate Assessment and Development Advisory Committee, draft *Third Climate Assessment Report*** (available at http://ncadac.globalchange.gov/). This draft report discussed how co- occurrence of heat waves and droughts can amplify impacts on water and electricity supply and demand, which can affect the energy sector in multiple and cascading ways. The report stated (p. 167 and p. 397, respectively) that "both episodic and long-lasting changes in water availability will constrain different forms of energy production," and "dependence of energy systems on land availability and water supplies will influence their development and constrain some options for reducing greenhouse gas emissions." It also noted (p. 387) that "jointly considering risks, vulnerabilities, and opportunities associated with energy, water and land use is difficult, but can improve the analysis of options for reducing climate change impacts." Recommendations for related research were included in a technical report developed to support the assessment effort (Pacific Northwest National Laboratory, *Climate and Energy-Water-Land System Interactions*, March 2012, http://www.pnnl.gov/main/publications/external/technical_ reports/PNNL-21185.pdf).
- **July 2013, Department of Energy (DOE), *U.S. Energy Sector Vulnerabilities to Climate Change and Extreme Weather*** (available at http://energy.gov/sites/prod/files/2013/07/f2/20130710-Energy-Sector-Vulnerabilities-Report.pdf). The report (p. i) stated the following when discussing the impacts of climate change on the energy sector and the potential for cascading and compounding impacts:

  > Some of these effects, such as higher temperatures of ambient water used for cooling, are projected to occur in all regions.

Other effects may vary more by region, and the vulnerabilities faced by various stakeholders may differ significantly depending on their specific exposure to the condition or event. However, regional variation does not imply regional isolation as energy systems have become increasingly interconnected. Compounding factors may create additional challenges. For example, combinations of persistent drought, extreme heat events, and wildfire may create short-term peaks in demand and diminish system flexibility and supply, which could limit the ability to respond to that demand.

The report identified a number of opportunities to enhance information, tools, and practices to reduce the energy sector's climate vulnerabilities. Some of the opportunities identified (p. 44) included better regional and local characterization of climate trends and extreme weather relevant to the energy sector (e.g., water availability, likelihood and magnitude of droughts); better characterization of the aggregate vulnerabilities of the energy sector to climate change and interdependencies with other sectors leading to cascading impacts; improved understanding of potential uses and challenges of advanced cooling technologies and alternative water sources; and additional assessments of impacts to hydropower.

- July 2013, Alliance for Water Efficiency and the American Council for an Energy-Efficient Economy, *Water-Energy Nexus Research: Recommendations for Future Opportunities* (available at http://www.allianceforwaterefficiency.org/WE-WhitePaper-PR.aspx). Its recommendations included continuing investigations into the water energy tradeoffs of differing resource development and management choices; identifying regulatory barriers to co-implementation of energy and water efficiency programs; developing water and energy industry-accepted protocols for efficiency programs; and assessing potential impacts to water supplies and quality from energy resource development and identifying solutions to mitigate these impacts.

Available Data on Water Use Remain Problematic

As of mid-2013, gaps remained in the availability of authoritative and recent data on the amounts of freshwater consumed and wastewater produced in fuel production. Available data indicate the following:

- The amount of water needed per unit of fuel produced—referred to as the water intensity of a fuel—ranges from conventional natural gas at the lowest end (less than 1 gallons of water per MMBtu);[12] coal, unconventional gas, and uranium mining and enrichment next (roughly 1 to 10 gallons per MMBtu); oil next (10 to 100 gallons per MMBtu);[13] and irrigated biofuels at the upper end (100 to 1,000 gallons per MMBtu).[14] The water intensity of conventional and unconventional oil produced using different techniques remains poorly documented. The water intensity for hydraulically fractured wells often is less notable than the concentrated, simultaneous demand for water for hydraulic fracturing in a region where many wells are being developed concurrently.
- Despite the recent increase in water demand for hydraulic fracturing, water use for stimulating oil production from conventional wells through water flooding and enhanced oil recovery have represented the largest water use by the oil and gas sector in the United States.[15] The use of these techniques is anticipated to increase; to what extent saline, wastewaters, or freshwater will be used is less clear. Limited data on production rates and quantities for many saline aquifers can be a disincentive to their use.
- Each fuel and production technique presents its own risks, potential water quality impacts,[16] and wastewater issues; also, some techniques may be more water- efficient but less efficient at recovering energy resources.[17] Data remain poor on the range of wastewater quantities and qualities derived from conventional and unconventional fuel production.

Fuel Production Remains Vulnerable to Water-Related Disruptions

The vulnerability of fuel production to freshwater availability is receiving attention in part because of increasing water demands (e.g., population growth) and concerns over changes to water supplies (e.g., drought and climate change).

- Instances of low flow and drought conditions have reduced the availability and increased the cost of water for operations in some locations (e.g., Susquehanna River basin in VA, WV, and PA, and Eagle Ford Shale region in TX). No analysis is available of the risk posed by a multi-year drought in areas of intense water use for energy (e.g., North Dakota) and how to manage the risk.
- Fossil fuel transport also may be disrupted by water conditions, such as flood- induced pipeline breaks resulting from riverbed scouring, flood- or storm-related refinery or distribution system disruptions (e.g., Hurricane Sandy disruptions), and drought- or flood-impaired fuel transport. No analysis of energy sector transport risks is available.

Wastewaters Represent Management Challenges and Some Opportunities

Produced water—wastewaters (often saline) brought to the surface by oil and gas wells— represents the largest byproduct of fuel production. Approximately 2.3 billion gallons are produced daily from onshore oil and gas wells in the United States.[18] For oil wells, this represents an average ratio of 7.6:1 of produced water to oil produced. By 2025, as a result of aging wells with decreasing oil production, the ratio is expected to average 12:1 for onshore crude oil.[19]

- U.S. energy-related wastewaters are primarily from conventional oil and natural gas and coal bed methane (CBM).[20] Research indicates that shale gas may produce less wastewater per unit of recovered gas than conventional natural gas.[21] Disposal of shale wastewaters has received more attention recently than wastewaters from conventional production because of the rate of increase in shale development and its associated wastewaters in locations that are not accustomed to oil and natural gas development.
- Management of energy-related wastewaters is evolving rapidly, with different techniques dominating in different locations and raising concerns related to water quality and seismicity. Where deep wells for the permanent disposal of produced water are limited, producers

increasingly are recycling and reusing produced water in fracturing operations. This reduces the amount of freshwater needed and relieves stress on disposal sites. At the same time, reuse of produced waters may increase the transport and handling of saline waters, potentially increasing a risk pathway for spills.

- The potential for human-induced seismic events associated with oil and gas wastewater injection is receiving scientific and political scrutiny in the context of the recent frequency of earthquakes, particularly in the central and eastern United States.[22] Over 300 earthquakes of magnitude (M) 3 or greater occurred between 2010 and 2012 in the central and eastern United States, compared to an average of 21 earthquakes per year of M>3 between 1967 and 2000.[23] The increase in seismicity seems to be correlated with the increase in the number of disposal wells, many of which are injecting wastewaters brought to the surface at shale oil and gas wells and in the use of hydraulic fracturing. Of the 30,000 injection wells in the United States used for wastewater disposal, however, only a few are correlated with seismicity of $M > 3$.[24] The largest seismic events associated with injection seem to involve faulting that is deeper than the wastewater injection, suggesting that transmission of pressure into the basement rocks elevates the potential for inducing earthquakes.[25]

Recent state actions and anticipated federal actions are affecting or are anticipated to affect the management of produced water.

- In Texas, produced water generally is disposed through deep-well injection (often on-site) or evaporation ponds; interest in reuse is increasing as the result of limited water availability in some regions (e.g., West Texas) and recent drought conditions. In May 2013, the Texas legislature clarified liability and ownership of produced waters transferred among oil and gas operators for purposes of recycling for a beneficial reuse.[26]
- Pennsylvania regulations constraining surface water disposal wastewaters from shale gas production and the limited in-state deep well-injection options have resulted in a rapid increase in the rate of produced water recycling for shale gas fracking.[27] Operators in Pennsylvania are required to prepare a wastewater source reduction strategy to maximize recycling and reuse.

- In August 2013, EPA proposed to discontinue efforts to establish discharge standards for wastewaters from CBM under the agency's Effluent Guidelines Program. EPA has been unable to identify a wastewater treatment technology that would be economically achievable.[28] The agency will continue with a rulemaking for wastewaters associated with shale gas extraction, which is expected to be proposed in 2014.

ELECTRIC GRID AND GENERATION

Water availability issues, such as regional drought, low flow, or intense competition for water, can curtail hydroelectric and thermoelectric generation. Fuel and power plant choices and capital investments made in the near term are likely to establish the trajectories for electric generation's long-term water use and vulnerability.

Grid-Level Drought Vulnerability Exists in Select Basins

An assessment of the drought vulnerability of electricity in the western United States found the majority of basins showing limited disruption risk; also, most of this risk could be mitigated by known strategies, including maintaining excess generation and transmission capacity.[29] While identifying broad resiliency, the western U.S. assessment revealed two regions whose electric generation was at greater risk:

- The Pacific Northwest was shown to be vulnerable because of its heavy reliance on hydroelectric generation.
- The Texas grid was vulnerable because of heavy dependence on thermoelectric generation that relied on surface water for cooling, and because of the region's vulnerability to drought and poor connections to the other U.S. grids, which reduces the ability to purchase power to offset generation curtailment.

No similar assessment of grid drought vulnerability for the eastern United States has been performed. (See following section, "Thermoelectric Cooling Represents Difficult Tradeoffs," for a discussion of electric generation in the eastern United States.)

Recent drought experiences include the following:

- In the summer of 2011, high temperatures in Texas resulted in increased electricity demand. At the same time, the drought reduced the amount of water available for cooling electric generators. The grid operator put into effect its emergency action alert system, which at first recommended conservation by customers and later deemed customer conservation critical to avoid rotating outages. During a few days, the peak demand purchases in the real-time wholesale electricity market were at or near the market cap (i.e., $3,000 per megawatt-hour). In the end, only one Texas plant had water-curtailed generation; others were nearing curtailment when weather conditions improved.[30]
- During the drought of 2012, the mid-continent electric grid avoided major drought-related disruption. Some individual power plants curtailed operations due to water access problems or water temperature issues; others pursued regulatory waivers to continue operations at higher water temperatures or made cooling system investments. Lost generation at drought-impaired facilities was offset by other generation or purchasing power from other sources on the wholesale market.

Hydropower Vulnerability Is Poorly Documented, But Data Are Expected to Improve

In Section 9505 of P.L. 111-11, Congress required the Secretary of Energy to assess the risks posed by climate change for water supply to federal hydroelectric power generators; the report is anticipated in 2013. Depending on the storage capacity (and its uses) in a basin's reservoirs, hydropower may be vulnerable to seasonal, annual, or multi-year drought conditions. For large reservoirs and reservoir systems, it is often the multi-year droughts that most harm generation,[31] as illustrated by summer 2013 conditions in the Colorado River Basin.

Thermoelectric Cooling Represents Difficult Tradeoffs

More than 80% of U.S. electricity is generated at thermoelectric facilities that depend on cooling water; these facilities withdraw 143 billion gallons of freshwater daily. The two common cooling methods for thermoelectric power plants are once-through cooling and evaporative cooling. Most once-through cooling is found at power plants located in the eastern United States and is associated with older facilities, or is at coastal facilities using saline waters. Newer facilities and those in more arid regions generally use evaporative cooling.

- Once-through cooling, while largely non-consumptive, requires water to be continuously available for power plant operations.[32] This reduces the ability for this water to be put toward other water uses and can make cooling operations vulnerable to low flows.
- Evaporative cooling withdraws much smaller volumes of water for use in a cooling tower or reservoir, where waste heat is dissipated by evaporating the cooling water. Evaporative cooling consumes more water at the facility than does once-through cooling.
- Cooling technologies that consume less water and use degraded water supplies may reduce freshwater use. These options include dry cooling, hybrid dry-wet cooling, cooling with fluids other than freshwater (e.g., brackish groundwater, produced waters), and emerging technologies. While hybrid and dry cooling options may reduce water consumption, they can reduce operational efficiency (potentially increasing greenhouse gas emissions) and often are more costly.[33]
- Future withdrawals associated with electric generation may grow slightly, remain steady, or decline depending on a number of factors, including reduced generation from facilities using once-through cooling (industry actions resulting from proposed federal cooling water intake regulations)[34] or shifts in how electricity is generated (e.g., less from coal and more from wind and natural gas).[35] In comparison, water consumption is more likely to increase, especially if more water-consumptive cooling is adopted (e.g., evaporative cooling) and if current carbon capture technologies are added to power plants.

Many Power Plants Produce Wastewaters

In addition to water for cooling purposes, many power plants also use water for handling solid waste, including ash, and for operating wet flue gas desulfurization scrubbers. According to the U.S. Environmental Protection Agency (EPA), in 2009, power plants discharged 0.7 billion gallons of wastewater daily. EPA recently proposed revisions to Clean Water Act rules that govern wastewater discharges from such plants. The proposed rule would reduce the use of these process waters by 19%-58%, depending on the regulatory option selected when the rule is finalized.[36]

POLICY RESPONSE OPTIONS AND CONSIDERATIONS

Policy makers at the federal, state, and local levels are deciding whether to respond to the growing water needs of the energy sector, and if so, which policy levers to use. In the United States, private entities make many of the energy sector's water decisions. Often federal entities lack authority over water use, and states have most of the water allocation authority. Instead of direct influence on water use, the public sector influences private water decisions through other routes (e.g., tax incentives, loan guarantees, permits, regulations, planning, and education). If action to manage energy-sector water issues is deemed appropriate, a range of options are available, as shown in **Table 1**: minimize water use, facilitate access to water, or improve decisions and data. Energy choices represent complex tradeoffs; water use and wastewater byproducts are two of many factors to consider. For many policy makers, concerns other than water—low-cost reliable energy, energy independence and security, climate change mitigation, public health, and job creation—are more significant drivers of their positions on energy policies.

Analyses quickly get complex when attempting to comprehensively evaluate energy-water tradeoffs. Some energy alternatives, such as solar photovoltaics and wind turbines, do not pose significant energy-water tradeoffs, but may pose other challenges, such as intermittent production or reduced dispatchability, which is the ability and ease with which output from an electric generation facility can be altered. Other energy tradeoffs include transport and storage. Some fuels are easier to store and use existing transport networks and multiple transport modes, while others may require new or expanded infrastructure investments (e.g., pipelines). Significantly, low-carbon energy is not necessarily low in water or environmental impact (e.g., new hydropower reservoirs, freshwater-cooled utility-scale solar), and specific carbon mitigation policies and actions may increase or decrease water consumption. Because of these complexities and the difficulty in comparing different types of impacts, analyses supporting decision-making are often incomplete. It is within this complex and confusing context that policy decisions that influence future energy and related water policies are being made.

Table 1. Policy Responses to Water Demands of Energy Sector

Water Demand Management Options	Water Supply Management Options	Options for Knowledge Development and Use
Minimize Energy Sector's Growth in Water Use	**Improve Energy Sector's Access to Water**	**Support Informed Decision-Making**
Promote water-efficient energy sources through standards, regulations, or incentives (e.g., rebates, water pricing)	Allocate sustainably available water, not otherwise allocated	Data and assessments; information sharing (e.g., data and research warehousing)
Promote water conservation and efficiency in the energy sector through standards, incentives, regulations, or pricing	Facilitate transfer of water from non-energy sectors (e.g., purchase of water from municipalities, or land owners; water markets)	Education, training, and dissemination of knowledge and information
Promote energy conservation and efficiency to reduce demand for energy and the embedded water		Integrated energy-water planning; coordination of research, decisions, and investments
Support research, development, scaling up, or adoption of technologies to reduce energy sector water use (e.g., public-private research collaborations)		Decision-support research and technical assistance; development of standard protocols and codes

Source: CRS.

End Notes

[1] In this report, production encompasses extraction and processing of fuels.

[2] This report complements CRS Report R41507, *Energy's Water Demand: Trends, Vulnerabilities, and Management*, by Nicole T. Carter, which analyses how and where the energy sector uses water in the United States.

[3] Consumption represents the water not available for immediate subsequent use. In the energy sector, water is consumed when it enters the atmosphere (e.g., power plant evaporative cooling towers), is lost to geologic formations, is sufficiently degraded to require permanent disposal, or needs treatment before use in freshwater applications or return to the environment.

[4] CRS Report R41507, *Energy's Water Demand: Trends, Vulnerabilities, and Management.*

[5] U.S. Geological Survey, *Estimated Use of Water in the United States in 2005* (Circular 1344: 2009).

[6] D. Elcock, "Future U.S. Water Consumption: The Role of Energy Production," *Journal of the American Water Resources Association*, vol. 46, no. 3 (June 2010), pp. 447-480, hereinafter referred to as Elcock 2010. Some estimates put thermoelectric water consumption closer to 3%. Available data is anticipated to improve when the U.S. Geological Survey releases its five-year water use survey for 2010; the agency has stated that it is resuming its estimates (which had stopped with the 1995 data) of water consumption by thermoelectric power plants and released a report in November 2013 on the improved methods that it will be using for determining those estimates.

[7] R. Young and E. Mackres, *Tackling the Nexus: Exemplary Programs that Save Both Energy and Water*, Washington, DC: American Council for an Energy-Efficient Economy, 2013.

[8] One study found that expanding the nation's electricity portfolio to 20% wind by 2030 would reduce water consumption by 1.2 billion gallons daily compared to expanding the current electricity mix. The water saved would be 41% in the Midwest/Great Plains, 29% in the West, 16% in the Southeast, and 14% in the Northeast (DOE, *20% Wind Energy by 2030: Increasing Wind Energy's Contribution to U.S. Electricity Supply*, July 2008, http://www1.eere.energy.gov/wind/pdfs/41869.pdf).

[9] Hydraulic fracturing is a technique developed initially to stimulate oil production from wells in declining oil reservoirs. The technique now is widely used to initiate oil and natural gas production in unconventional (low-permeability) formations that were previously inaccessible. Fracturing is currently used in more than 90% of new oil and natural gas wells. Hydraulic fracturing involves injecting large volumes of water, sand (or other propping agent), and specialized chemicals under pressure into a well to fracture the formations holding trapped oil or natural gas.

[10] U.S. Government Accountability Office (GAO), *Energy-Water Nexus: Coordinated Federal Approach Needed to better Manage Energy and Water Tradeoffs*, GAO-12-880, September 2012, http://www.gao.gov/assets/650/ 648306.pdf.

[11] The site links to over 150 items related to energy-water issues: http://en.openei.org/wiki/Water_and_energy_studies. Also the Energy Information Agency in recent years has increased the type and frequency of data collection on power plant cooling systems. More state level data is being collected; for example, the Railroad Commission of Texas required oil and natural gas operators to disclose on FracFocus (http://fracfocus.org/) water volumes and chemicals used for hydraulic fracturing after February 2012.

[12] MMBtu represent 1 million British thermal units which is a commonly used unit of energy.

[13] Oil is produced by a variety of techniques, some of which can be particularly water-intensive (e.g., water flooding). Oil shale is largely not discussed herein. Oil shale is distinct from the tight oil produced from shale formations. Oil shale's near-term impacts on water resources are limited by the relatively small scope of anticipated near-term development (GAO,

Unconventional Oil and Gas Production: Opportunities and Challenges of Oil Shale Development, Washington, D.C, 2012).

[14] International Energy Agency, *World Energy Outlook 2012*, Paris, 2012; E. Mielke, et al., *Water Consumption of Energy Resource Extraction, Processing, and Conversion*, Cambridge, Massachusetts: Harvard Kennedy School, 2010; World Energy Council, *Water for Energy*, London: World Energy Council, 2010.

[15] M. Matichich, *The Changing Value of Water to the US Economy: Implications from Five Industrial Sectors*, Boston: CH2M Hill, 2012.

[16] A discussion of water quality impacts is beyond the scope of this report. For a regional discussion of water quality concerns associated with shale gas, see CRS Report R42333, *Marcellus Shale Gas: Development Potential and Water Management Issues and Laws*, by Mary Tiemann et al. For a general discussion, see L. Allen, et al., Fossil Fuels and Water Quality, in P. Gleick, *The World's Water*. Vol. 7, Washington, DC: Island Press, 2012, pp. 73-96.

[17] Beyond water considerations, fuel production can have other development impacts (e.g., roads, housing). For example, see CRS Report R42611, *Oil Sands and the Keystone XL Pipeline: Background and Selected Environmental Issues*, coordinated by Jonathan L. Ramseur.

[18] This compares to an estimated 4.6 billion gallons per day of freshwater used for fuel production.

[19] Global Water Intelligence, Water's growing role in oil and gas, March 2011, http://www.globalwaterintel.com/archive/12/3/market-profile/waters-growing-role-oil-and-gas.html.

[20] CBM production generally requires the dewatering of a coal formation for the natural gas to be released; the quantity and quality of CBM produced waters varies widely across formations (e.g., salinities ranging from freshwater or more saline than seawater). For more on CBM water issues, see National Research Council, *Management and Effects of Coalbed Methane Produced Water in the Western United States*, National Academies Press, August 2010.

[21] B. Lutz, A. Lewis, and M. Doyle, "Generation, transport, and disposal of wastewater associated with Marcellus Shale gas development," *Water Resources Research*, 49 (2013).

[22] William I. Ellsworth, "Injection-Induced Earthquakes," *Science*, vol. 341 (July 12, 2013).

[23] Ibid., p. 1225942-1.

[24] Ibid., p. 1225942-2.

[25] Ibid., p. 1225942-6.

[26] H.B. 2767 (Texas). Also in 2013, the Railroad Commission of Texas stopped requiring a recycling permit if operators are recycling on their own leases or transferring fluids to another operator's lease for recycling ("Railroad Commission Today Adopts New Recycling Rules to Help Enhance Water Conservation By Oil & Gas Operators," press release, March 26, 2013).

[27] J. Logan, et al., *Natural Gas and the Transformation of the U.S. Energy Sector: Electricity*. Golden, Colorado: Joint Institute of Strategic Energy Analysis, 2012, http://www.nrel.gov/docs/fy13osti/55538.pdf.

[28] EPA, *Economic Analysis for Existing and New Projects in the Coalbed Methane Industry*, EPA 820-R-13-006, July 29, 2013, http://water.epa.gov/scitech/wastetech/guide/oilandgas/unconv.cfm

[29] C.B. Harto, et al., *Analysis of Drought Impacts on Electricity Production in the Western and Texas Interconnections of the United States*, Oak Ridge, Tennessee: U.S. Department of Energy, 2011.

[30] During and after the summer of 2011, Texas power plant operators reduced their low water vulnerability by building pipelines to alternative and impaired water sources, acquiring additional water rights, lowering water intake structures, and installing additional groundwater pumping capacity. Also, the Texas grid operator instituted changes to reduce its water vulnerability. All new generation facilities as of 2013 must provide proof of water rights before being included in grid planning (which largely determines grid access). Few

data are available on the extent to which low-water renewable technologies may be used to mitigate the Texas grid's drought risks.

[31] For example, in 2012, hydropower production nationally was above average although drought conditions covered much of the continental United States. The Missouri River basin's strong hydropower generation in 2012 can be attributed to full reservoirs at the beginning of the year and the generation associated with releases of stored water to augment low river flows.

[32] Once-through cooling pulls large quantities of water off a water body, discharges the power plant's waste heat into the water (which may raise the temperature of the withdrawn water by 10° to 20°F), and then returns the majority of the withdrawn water.

[33] The National Science Foundation and the Electric Power Research Institute have an ongoing research collaboration on water for energy to further advanced dry cooling for power plants.

[34] See CRS Report R41786, *Cooling Water Intake Structures: Summary of EPA's Proposed Rule*, by Claudia Copeland.

[35] Natural gas-fueled generation is often less water-intense and less water-dependent than coal-powered electricity. This is because many natural gas-fueled electric facilities use engine-based technology (National Energy Technology Laboratory, *Estimating Freshwater Needs to Meet Future Thermoelectric Generation Requirements* (2009 Update), 2009).

[36] EPA, *Technical Development Document for the Proposed Effluent Limitations Guidelines and Standards for the Steam Electric Power Generating Point Source Category*, EPA-821-R-13-002, April 2013, pp. 12-13 – 12-14.

In: Energy and Water Sector Interdependence ISBN: 978-1-63117-789-7
Editor: Wilfred D. Meadow

Chapter 3

ENERGY'S WATER DEMAND: TRENDS, VULNERABILITIES, AND MANAGEMENT*

Nicole T. Carter

SUMMARY

The energy choices before Congress represent vastly different demands on domestic freshwater because water is used in varying amounts in most aspects of the energy sector. Transitions in the energy sector, such as the pursuit of greater energy independence and security, produce changes in how much and where the energy sector uses water. The energy sector is the fastest-growing water consumer in the United States, in part because of energy policies. Whether the federal government addresses the energy sector's rising water demand, and if so how, is one of the many energy decisions that may be considered by the 112th Congress.

Much of the growth in the energy sector's water demand is concentrated in regions with already intense competition over water. Whether the energy sector exacerbates or alleviates future water tensions is influenced by near-term energy policy and investment decisions. These decisions also may determine whether water will limit or harm U.S. capability to reliably meet the nation's energy demand. Part of the policy

* This is an edited, reformatted and augmented version of Congressional Research Service, Publication No. R41507, dated January 5, 2011.

issue for Congress is identifying the extent of the federal role in responding to the energy sector's water use. Currently, the energy industry and states have the most responsibility for managing and meeting the energy sector's water demand.

The energy sector's water consumption is projected to rise 50% from 2005 to 2030. Projections attribute to the energy sector 85% of the growth in domestic water consumption between 2005 and 2030. The drivers of the energy sector's increasing water use are rising energy demand, greater development of domestic energy, and shifts to more water-intense energy sources and technologies. The more water used by the energy sector, the more vulnerable the energy system is to water availability. Climate change alterations of historic water patterns may exacerbate this vulnerability in some regions. While the energy sector's water demand is anticipated to rise across the United States, the West is likely to experience some of the more significant constraints in meeting this demand. Examples of regional water use concerns related to energy are shale gas production using hydraulic fracturing in many regions across the nation, some solar electricity generation in the Southwest, and current biofuel feedstock production in the High Plains.

The 112th Congress may see the issue of energy's water demand in a variety of contexts, including oversight and legislation on energy, infrastructure, environment, agriculture, public lands, climate, research, and water. Approaches for addressing energy's water demand range from maintaining the current approach to taking a range of actions. One set of available actions includes those that attempt to minimize the growth in energy's freshwater use (e.g., through promotion of water-efficient energy alternatives and energy demand management), which could be accomplished through changes to broad policies or legislation targeted at water use. Many of the possible actions to decrease water use come with energy cost, generation, and reliability penalties. Another set of actions includes measures that facilitate access to water for the energy sector. While water allocations and permits generally are a state responsibility, limited federal actions to provide water to the energy sector are possible (e.g., access to surplus water at federal reservoirs). An additional set of actions encompasses investments in data and research to inform decision making and to expand water-efficient energy technology choices. These approaches represent different potential roles and expenses for government, the energy sector, and energy consumers. Legislation in the 111th Congress proposed a variety of actions, including provisions in H.R. 469, H.R. 2454, H.R. 3598, H.R. 1145, S. 1462, S. 1733, S. 3396, and the Secure Water Act of 2009, Subtitle IV of P.L. 111-11 (H.R. 146). A significant challenge to a federal response to the energy sector's water demand is that the available options are not equally needed, attractive, or feasible across the United States.

ENERGY'S WATER USE: A POLICY INTRODUCTION

The nation's energy choices embody many tradeoffs. Water use is one of those tradeoffs. The energy sector is the fastest-growing water consumer in the United States. Projections attribute to the energy sector 85% of the growth in domestic water consumption between 2005 and 2030. This projected growth derives from anticipated demand for more energy, increased development of domestic energy sources, and greater use of water-intense energy alternatives. Much of the energy sector's growing water demand is concentrated in water-constrained regions. Affordable water supplies are limited, and competition for water is becoming more intense. Whether the energy sector helps exacerbate or alleviate future water tensions is influenced by current energy policy and investment choices. These choices also may determine whether water limits or harms the nation's ability to reliably meet energy demand. Water limitations may hinder some water-dependent energy activities in specific locations.

Water already plays a significant role in the energy sector, and water use by the energy sector already shapes national water use. For example, more than 80% of U.S. electricity is generated at thermoelectric facilities. With few exceptions, these thermoelectric power plants are cooled with water.[1] In 2005, withdrawal of water for cooling represented 44% of water withdrawn nationally,[2] and 6% of the water consumed nationally.[3] The more water used by the energy sector, the more vulnerable energy production and reliability are to competition with other water uses and water constraints such as droughts. Climate change impacts on water supplies in some regions may exacerbate this vulnerability. Water availability can affect both existing and new energy activities, as well as all those economic activities that depend on the fuels and electricity produced.

The energy sector is changing. Paths chosen and capital investments made in the near term are likely to establish long-term trajectories for energy's water use. Trends indicate that energy's changing water use has national and regional significance for water consumption. A question for Congress is: what is the appropriate federal role in responding to energy's water demand? In the aggregate, current federal energy policies contribute to energy's rising water demand, while energy interests and the state and local governments are responsible for managing and meeting water demands and resolving competition over water resources.

Questions for Congress include *who* is the most appropriate entity to respond to energy's growing water demand and water vulnerability and *how* to respond. At present, little direct federal action is aimed at managing the energy sector's water demand; instead, the current division of responsibilities relies on energy interests and state and local governments to meet and manage energy's water demand and resolve energy-water conflicts. The role of federal policies in contributing to rising water demand is bringing into question the future federal role in this policy arena. Local or regional competition for water with existing users is often what makes energy's water demand significant; at the same time, the regional and local scales of water resources availability and management complicate many federal water-related actions.

Options for managing and meeting energy's water demand range from maintaining the current approach, with little federal action targeted at managing energy's water demand, to taking a variety of federal actions. One option is to minimize growth in energy's freshwater use. This could be accomplished through changes to broad policies (e.g., energy demand management) or legislation specifically targeted at water use (e.g., promotion of water-efficient energy alternatives). Another option is to improve the energy sector's access to water. Access is generally a responsibility of the state, but some limited federal actions are possible. An additional option is investing in data and research to inform decision-making and expand water-efficient energy technologies. These alternative policy approaches, which are not mutually exclusive, represent different potential roles and costs for the energy sector; energy consumers; and federal, state, and local governments. Legislation in the 111th Congress proposed many of the above options; examples include H.R. 469, H.R. 2454, H.R. 3598, H.R. 1145, S. 1462, S. 1733, S. 3396, and Subtitle IV of P.L. 111-11 (H.R. 146), the Secure Water Act of 2009.

During the 112th Congress, energy's water use may arise in a variety of contexts, including during consideration of energy, agriculture, public land, and water legislation and oversight. While the water tradeoffs of energy choices may be raised in a variety of contexts, they are unlikely to be the focus of energy debates. Instead, the priority on and investments toward different policy goals—low-cost reliable energy, energy independence and security, climate change mitigation, and job creation—are likely to be more significant drivers in congressional energy deliberations.

Scope and Structure of This Report

This report focuses on the factors shaping the energy sector's water demand, how that demand fits into national water consumption, and options for managing and meeting water demand. The report first lays out the trends shaping energy's water use; second, it discusses energy's vulnerability to water constraints; and third, it discusses projections of energy's water use. It then explores three regional examples of energy's water use: shale gas in Texas, solar energy in the Southwest, and biofuels in the High Plains. Finally, it discusses policy options and legislative approaches for managing energy's water use. Several appendixes provide more detailed information on specific technologies, fuels, and trends.

The report does not discuss in detail the energy sector's water quality impacts, although they represent their own challenges, as shown by concerns over the water quality effects of hydraulic fracturing, mountaintop mining,[4] and the Deepwater Horizon oil spill.[5]

Energy use by the water sector also is not discussed in this report, although water conservation is one of many available means for reducing energy demand.

The Federal Role in Water Allocation, Management, and Planning

The United States is endowed with considerable water resources; even in the drier West there are some major rivers and substantial groundwater supplies. Responsibility for development, management, and allocation of the nation's water resources is spread among federal, state, local, tribal, and private interests.

Water Allocation

The states administer most water rights allocation, deciding how to distribute freshwater among users. The mechanisms for distribution differ for each state, with states in the East and West historically following two different systems of water law. Distribution becomes more contentious as local or regional demand and environmental needs outpace sustainable and affordable supplies. States in large part decide whether and how to adapt allocation mechanisms and institutions to meet changing demands and priorities, and whether to allow or facilitate the movement of water among alternate uses. However, the federal role in allocation increases as the ramifications of state water allocations are felt in other states or internationally, or when they run contrary to federal law, such as the Endangered Species Act or the Clean Water Act.

Rather than directly influencing state allocation laws, decisions, and institutions, the federal government usually uses indirect means to influence water use, such as programs for agricultural water conservation, investments in water augmentation research, and support for water resource planning.

Water Management

The federal government has been called upon to assist with and pay for a multitude of water resource development projects. Federal works range from improvements to facilitate navigation beginning in the earliest days of the nation, to more recent efforts to reduce flood damages and expand irrigation. In recent decades, the federal government also has regulated water quality, protected fish and wildlife, and facilitated some water supply augmentation. Criticism of the fractured nature of federal water policy and concerns about the efficiency of current water use have been recurrent themes for decades. Congress has not enacted comprehensive changes in federal water resources management or national water policy since enactment of the 1965 Water Resources Planning Act (P.L. 89-80; 42 U.S.C. § 1962). While many stakeholders call for better coordination and a clearer national "vision" for water management, Congress often has reacted to such proposals as attempts to exert federal control over state and local matters or as attempts to concentrate power in the executive branch. Instead, Congress has enacted numerous incremental changes, agency by agency, statute by statute. Both the executive and judicial branches have responded to these changes and, over time, have developed policy and planning mechanisms on an ad hoc basis. When coordination of federal activity has occurred, it has been driven largely by pending crises, such as potential threatened or endangered species listings, droughts, floods, and hurricanes; and by local or regional initiatives. Concern about water supply, however, has bolstered recent interest in legislation to establish a national water commission and strategy.

Water Planning

Energy's rising water demand and the potential for climate change to decrease water supply availability and reliability in some regions are part of a recent interest in water planning.

Consensus, however, does not exist about the utility of or proper federal role in planning. Following the 1965 Water Resources Planning Act, the federal government supported federal, state, and river basin planning. By the late 1970s, this planning was both positively received and criticized for its costs and usefulness. The challenge of water planning is summed up in a 1973 report:

> A persistent tendency of water resources planning has been the issuance of single valued projections of water use into the future under a continuation of present policies, leading to astronomical estimates of future water requirements.... The amount of water that is actually used in the future will depend in large measure on public policies that are adopted. The National Water Commission is convinced that there are few water "requirements."... But there are "demands" for water and water-related services that are affected by a whole host of other factors and policy decisions, some in fields far removed from what is generally considered to be water policy. (National Water Commission, *Water Policies for the Future: Final Report to the President and to the Congress of the United States* (Washington: GPO, 1973), p. 2)

In the late 1970s and early 1980s, federal funding for state planning declined, and federal involvement shrank with the defunding of the executive-level Water Resources Council and most federal river basin commissions. Some water resources stakeholders continue to view federal involvement in planning as infringing on state primacy, while other stakeholders support greater federal involvement in watershed, multi-objective, or integrated planning efforts. Some states, such as California, Texas, and Florida, have undertaken their own planning efforts.

ENERGY TRENDS SHAPE WATER DEMAND

Trends in national energy investments, domestic energy use, population, and climate change impacts and responses can affect *how much* and *where* the energy sector uses water. Increased emphasis on domestic energy production and efforts to meet increasing energy demand are expected to increase freshwater use by the energy sector. A shift in the electricity sector away from traditional coal power plants may result in either more or less water consumption, depending on alternative fuels or electricity technologies. Carbon capture and sequestration (CCS) by electric utilities has the potential to consume significant quantities of water.[6] Actions such as substituting wind for thermoelectric electricity generation could potentially reduce energy's water demand, but may raise other challenges for energy reliability, dispatchability, and transmission. Other impacts, such as the movement of irrigated agriculture from food crops to energy crops, raise other concerns. These and other examples of energy trends and their effects on energy's water use are summarized in Table 1. Appendix A has a more extensive list of trends.

Table 1. Energy Trends Produce Water Use Trends

Energy Trend	Resulting Trend in Energy's Water Use
Shift from foreign oil to biofuels	Increases energy's water consumption if domestic agricultural irrigation water (and other inputs) is needed for fuel production.
Shift to shale gas	Natural gas development using hydraulic fracturing may raise water quantity concerns if well development is geographically concentrated in areas with water constraints. However, natural gas from fracturing consumes less U.S. freshwater than domestic ethanol or onshore oil.
Growth in domestic electricity demand	More water used for electricity generation; how much more depends on how the electricity is produced (e.g., smaller quantities needed if electricity demands are met with wind and photovoltaic solar, larger quantities if met with fossil fuels or certain renewable sources).
Shift to renewable electricity	Concentrating solar power technologies can use more water to produce electricity than coal or natural gas; these solar facilities are likely to be concentrated in water-constrained areas. Technologies are available to reduce this water use. Other renewable technologies, such as photovoltaic solar and wind, use little water.
Use of carbon mitigation measures	Carbon capture and sequestration may double water consumption for fossil fuel electric generation.

Source: CRS.

Whether and how much the energy sector's water demand grows in the next decades will be significantly influenced by whether energy demand increases. Projections of the size and mix of the future energy portfolio vary widely. These projections are highly uncertain and are sensitive to many factors, including market and economic conditions, energy and agricultural policies, resource availability, technology developments, and environmental regulations. By association, projections of energy's water demand also are highly uncertain.

The Energy Information Administration (EIA) in the Department of Energy (DOE) projects that the United States will consume 22% more electricity and 12% more liquid fuel in 2030 than in 2010.[7] Population growth and increased electricity use per capita are some drivers of increasing demand.

Significant shifts to more water-intense electricity generation (e.g., concentrating solar power facilities using evaporative cooling) or more water-intense fuels (e.g., oil shale) could increase energy's water demand in locations with these energy resources.[8]

The significance of energy's water demand depends in part on local conditions—how much water is locally available and what its alternative uses would be.

The most growth in freshwater consumption by the energy sector is expected in the Southwest, the Northwest, and the High Plains—that is, regions already experiencing intense competition over water and disputes over river and aquifer management.

Transportation's Water Consumption Is Increasing

The transportation sector offers an example of how energy trends affect water use. Water consumption by transportation fuels is anticipated to increase between 2005 and 2030; this increase is shaped by multiple trends.

As shown in the figure below, an increase in miles driven and the increasing water-intensity of fuels, as a result of irrigated biofuels (i.e., biofuels derived from irrigated feedstock), overwhelms the water gains from improving vehicle fuel efficiency. (Appendix B has more information on water consumed for a variety of transportation fuels.)

For light-duty vehicles (LDVs), the Energy Information Administration (EIA) projects a 50% increase in miles traveled, while the water-intensity of those miles is projected to rise from 40 gallons per 100 miles traveled to almost 90 gallons, according to researchers. Consequently, water consumption for LDV travel is roughly projected to triple from 2005 to 2030.

Water consumption for LDV travel increased 50% between 2005 and 2010.

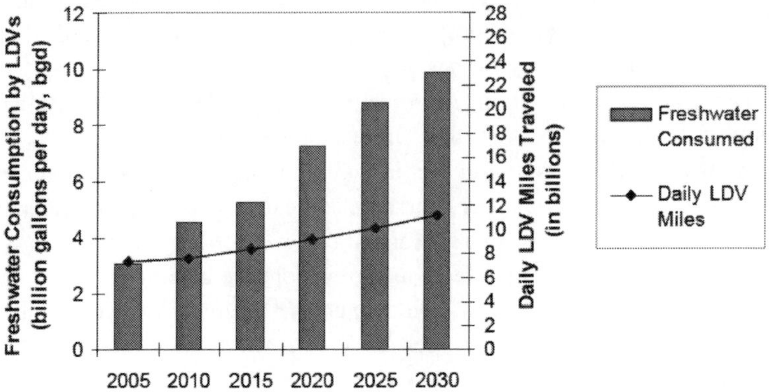

Sources: C. W. King, M. E. Webber, and I. J. Duncan, "The Water Needs for LDV Transportation in the United States," *Energy Policy*, vol. 38 (2010), pp. 1157-1167; EIA, *Annual Energy Outlook 2008*, Table 7, Transportation Sector Key Indicators and Delivered Energy Consumption, Line 15. The King 2010 article used 2008 EIA data.

ENERGY SECTOR'S VULNERABILITY TO WATER CONSTRAINTS

The more freshwater used by the energy sector, the more the sector is vulnerable to water constraints. However, as described above and in Appendix A, major energy trends are pushing the sector to become more water-intensive. Water availability problems, especially regional drought and low streamflow, can pose a risk to energy production and reliability. Electricity generation is particularly sensitive to low-flow conditions. More than 80% of U.S. electricity is generated at thermoelectric facilities that depend on access to cooling water. Low-flow conditions and water scarcity may constrain water-intense alternatives for thermoelectric cooling in counties across the country. Additional ways that water can constrain energy include a possible decrease in hydroelectric generation during drought. Bioenergy yields may be reduced by low precipitation, droughts, heat waves, or floods. Energy extraction, like coal mining, may be scaled back to avoid water quality impairments exacerbated by low water conditions. While water constraints are often perceived as an issue for the western United States, an increasing number of water bodies in the East are experiencing diminished stream flows.[9] While multiple examples exist of water availability affecting siting and operations of thermoelectric facilities from New York to Arizona,[10] generally there are ways to reduce the use of water and the risk posed by water constraints.

Water supplies often are most constrained during summer, when the energy sector's water use is at its height in many regions. Approximately 24 of the nation's 104 nuclear reactors are situated in drought-prone regions.[11] The Nuclear Regulatory Commission sets minimum source water elevation levels for each plant, so that a plant does not operate when source water levels drop below plant cooling water intakes. If cooling water sources fall below the established minimum water level, or if the maximum thermal threshold for the discharge of cooling water cannot be met, a facility is required to power down or go offline. A commonly cited example of this occurred on August 16, 2007, when a nuclear reactor owned by the Tennessee Valley Authority (TVA) at the Browns Ferry Nuclear Power Plant in Alabama shut down for a day. Its cooling water discharge exceeded temperature regulations that protect the environment and wildlife of the Tennessee River. In the summer of 2010, the same plant cut its electricity production to 45% of capacity when the cooling water temperature again exceeded discharge regulations.[12] The reduced generation resulted in $50 million in higher cost to customers. TVA subsequently initiated a $160 million upgrade and expansion of its cooling system to avoid similar cooling constraints in the future.[13]

Water Constraints on Thermoelectric Cooling

More than 80% of U.S. electricity is generated at thermoelectric facilities. Thermoelectric facilities generally can be used to produce power as needed, according to consumer demand and fuel supply. This responsiveness to demand makes electricity from these thermoelectric facilities particularly attractive.

Thermoelectric facilities can be fueled by a variety of fuels; coal, nuclear, and natural gas are the most common. Renewable sources such as concentrating solar power (CSP), geothermal, and renewable biomass also use a thermoelectric steam cycle. Thermoelectric power plants use fuel to produce heat to generate steam, which turns a turbine connected to a generator that produces electricity.

Cooling is required to condense the steam back into boiler feed water, so the process can be repeated. With few exceptions, water is used to cool existing U.S. thermoelectric power plants. Thermoelectric cooling represents 44% of the freshwater withdrawn nationally, and roughly 6% of the water consumed nationally.

The cooling options available for thermoelectric plants vary in their water withdrawal and consumption. Water withdrawal is the volume of water removed from a water source. Consumption is the volume lost, that is, no longer available for use. Excessive withdrawals can harm aquatic ecosystems, while excessive consumption depletes the water available for other uses. The two common cooling methods are once-through cooling and evaporative cooling. Once-through cooling pulls large quantities of water off a water body, discharges the power plant's waste heat into the water (which typically raises its temperature 10° to 20°F), then returns the majority of the withdrawn water.

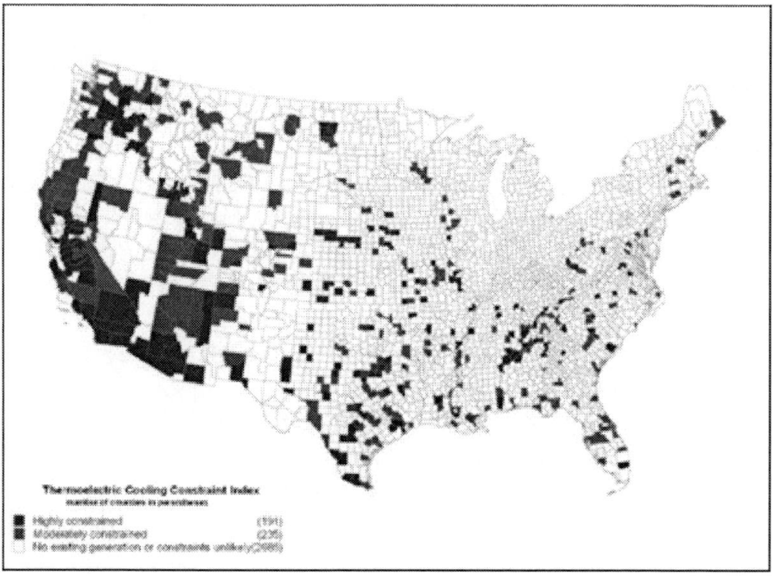

Source: EPRI, A Survey of Water Use and Sustainability in the United States with a Focus on Power Generation, Topical Report, Nov. 2003. EPRI's analysis did not include Alaska and Hawaii.

Once-through cooling, while largely a non-consumptive water use, requires that water be continuously available for power plant operations. This reduces the ability for this water to be put toward other water uses and can make cooling operations vulnerable to low streamflows. Evaporative cooling withdraws much smaller volumes of water for use in a cooling tower or reservoir, where waste heat is dissipated by evaporating the cooling water.

Evaporative cooling consumes water. Many power plants operating in the East use once-through cooling, while the majority in the West use evaporative cooling, although some coastal facilities use saline water for once-through cooling. In general, older thermoelectric plants use once-through cooling. The withdrawal's effect on the ecology and quality of the water body (e.g., elevated temperature and chemicals of the discharged cooling water) in once-through cooling have resulted in newer power plants generally using evaporative cooling. The figure below was produced by Electric Power Research Institute (EPRI) in 2003. It projects the counties where thermoelectric cooling may be constrained as the result of water availability in 2025; this figure was created assuming historic water availability; that is, it did not account for potential climate change effects on water supply or electricity demand.

Climate Change Could Increase Energy's Freshwater Vulnerability

Snowpack, precipitation, and runoff are strongly related to climate. Climate change researchers predict both water quantity and timing changes. That is, the research indicates more precipitation in the form of rain and less in the form of snow, and changes to seasonal water availability in some areas (e.g., low-flows during dry seasons). Additionally, climate models predict more frequent floods and droughts. These changes present challenges for hydroelectric dam operation.[14]

Changes in the availability and temperature of water resources also may affect operations of power plants that require water for cooling and that have thermal discharge limitations for cooling water. Climate change also may increase the demand for air conditioning, the electricity it consumes, and the water used to produce the electricity. The decreased runoff anticipated in the West, Southwest, California, and Pacific Northwest[15] would decrease the amount of water available for all uses, including the energy sector.

That is, the water resource impacts of a changing climate would likely exacerbate already projected thermoelectric cooling constraints. The energy sector also is vulnerable to potential increased flood and storm hazards associated with climate change. An example is the disrupting effects of floods on fuel transport.[16]

How Much Water Does Energy Demand?

How much water will the energy sector use in the future? In part, interest in answering this question is rooted in other questions of national significance, such as: Will water limit U.S. capacity to meet the nation's energy demand? Will water constrain the transition to greater energy independence? Will water hamper adoption of some renewable energy alternatives?

Quantification of energy's water demand and its significance is limited by significant gaps in available data and analyses. Water has no federal data agency comparable to the Energy Information Administration that projects alternative demand scenarios.[17] There is no authoritative government source to cite for the level of water use by the energy sector or for projections of how that use may change in future decades. For example, there are no forecasts that use multiple scenarios to identify sensitivity of water demand to multiple factors and policies, or that analyze energy's water use and water vulnerability in the context of factors significant to energy choices and policies, such as energy and transmission costs, emissions, and reliability.[18]

For over 50 years, the U.S. Geological Survey (USGS) of the Department of the Interior has collected and published water use data every five years; however, the agency stopped collecting water consumption data after its 1995 survey due to funding constraints and data reliability problems. (USGS continues to collect water withdrawal data.) The 1995 USGS data are the basis of most projections of future water consumption in the United States.[19] In the Secure Water Act of 2009, Congress authorized the USGS to perform a water use and availability assessment that includes water use trends in the energy sector; however, to date the agency had received no congressional appropriations for this program. Every 10 years the Forest Service forecasts water resources trends based largely on extrapolations of the USGS data and using the USGS data categories.[20] The energy sector falls into a number of USGS water use categories, and it is impossible to disaggregate the categories to determine a value for the energy sector's water use. (Except for thermoelectric water withdrawals, the USGS water use data and Forest Service projections do not break out energy water use from other agricultural and industrial water uses.) Because of these limitations, the analysis herein relies on the most comprehensive projections available on the energy sector's water consumption, with the main source being a 2010 article by Deborah Elcock, an Argonne National Laboratory researcher, based on an updated and refined analysis from a report published by the lab in 2008.[21]

Although currently available data are limited, there are prospects for improved data and analysis in the future. The Secretary of the Interior announced in October 2010 that the USGS is undertaking a Colorado River Basin Geographic focus study as part of the department's WaterSMART Water Availability and Use Assessments initiative; this focus study may eventually comprise a component of the USGS water use and availability assessment Congress authorized in 2009, which would represent the first national water census since 1978.[22] Additionally, regional efforts may inform future decision-making. For instance, the Western Governors' Association initiated in 2010 an energy-water nexus project, as part of its renewable energy transmission expansion effort, which includes a water availability assessment.[23] The assessment is looking into projected water demands for large river basins and aquifer systems in the West and is expected to consider drought and potential climate change implications on the availability of river flows and water supply for energy development in the West. The effort is anticipated to conclude with policy recommendations available in late 2012, and is funded primarily by DOE.

Energy Leads Projections of Increasing U.S. Water Consumption

Energy's Water Demand May Increase 50% from 2005 to 2030

During the 1980s and 1990s, as a result of improved water efficiencies, U.S. water consumption remained below the historic high level of 101 billion gallons per day (bgd) estimated for 1980.[24] Estimates of recent consumption, however, put U.S. freshwater consumption above the previous high and predict it will increase further in the next decades. (See Figure 1.) The projected rise is dominated by energy's water use. Nationally, energy's water consumption exceeds municipal and industrial use; it is currently second only to agriculture.

By modeling current energy and water trends, available projections predict that water consumption by 2030 will increase by 7% above the level consumed in 2005, as shown in Figure 1. Eighty-five percent of this growth is attributed to the energy sector, and an increase has already been observed between 2005 and 2010, with greater energy sector water use for irrigated biofuels.[25] (See Appendix B.) Energy's water consumption (excluding hydropower) in 2005 was roughly 12 bgd, as shown in Figure 2; it is estimated to increase to 18 bgd by 2030. As shown by the data summarized in Figure 2, roughly 60% of the anticipated expansion is associated with bioenergy, and 40% is associated with thermoelectric cooling and fossil fuel mining, production, and processing. The analyses behind these projections do not capture the effects on water consumption of various fuel and generation technology changes in the electricity sector and potential adoption of CCS, nor do they include hydropower's water consumption.

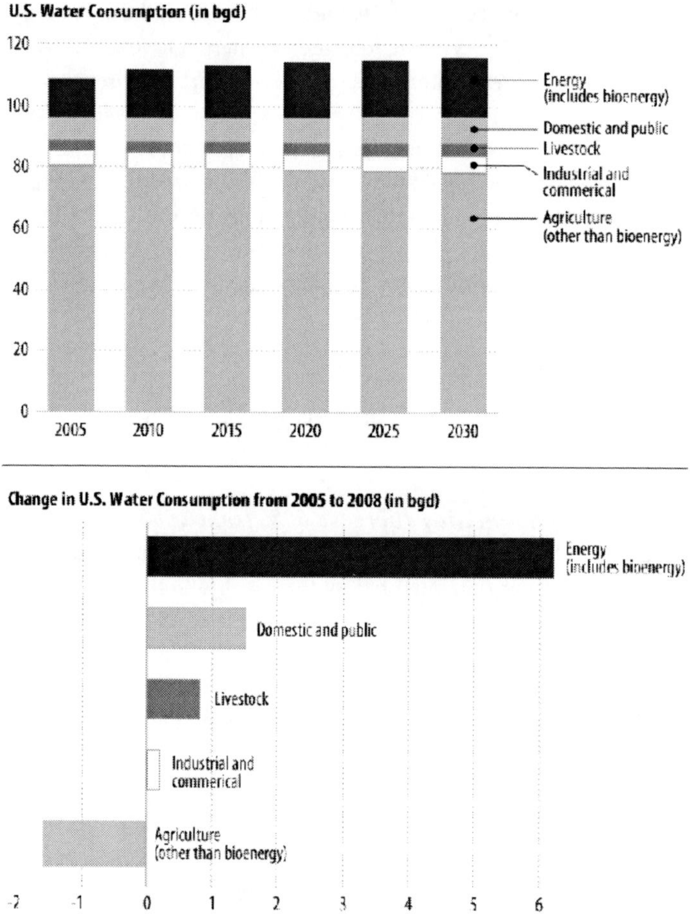

Source: CRS, using data from D. Elcock, "Future U.S. Water Consumption: The Role of Energy Production," *Journal of the American Water Resources Association*, vol. 46, no. 3 (June 2010), pp. 447-480.

Figure 1. Projection of U.S.Water Consumption.

Although the total national increase in Figure 1 may not appear large (7% over 25 years), multiple factors make this a significant increase for the water sector. First, growth in water consumption between 2005 and 2010 has put current water consumption above the previous high set in 1980. Second, water in many basins is largely allocated (or even over-allocated), with the water being delivered under legally binding agreements or withdrawn under issued permits. This means that making water available to the energy sector would likely decrease water use in another sector, such as agriculture. Third, Figure 1 represents national freshwater use; local proportions and increases in water demand from energy may be significantly higher.

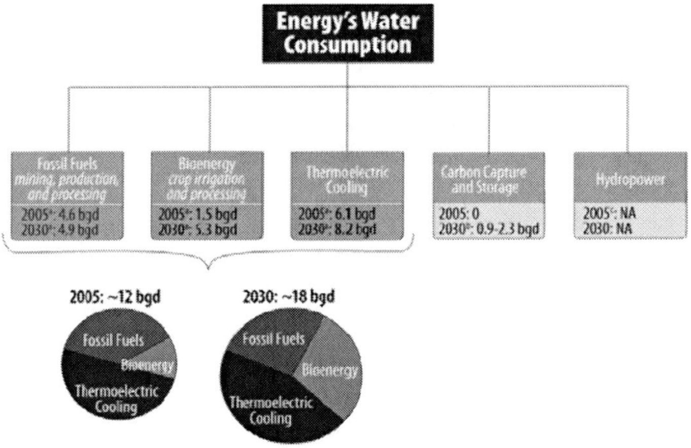

Source: CRS, using data sources noted below. Notes: NA = not available.

a D. Elcock, "Future U.S. Water Consumption: The Role of Energy Production," *Journal of the American Water Resources Association*, vol. 46, no. 3 (June 2010), pp. 447-480.

b NETL, *Estimating Freshwater Needs to Meet Future Thermoelectric Generation Requirements (2009 Update)*, p. 64.

c A 2003 National Renewable Energy laboratory (NREL) report estimated that evaporation at reservoirs at the 120 largest existing U.S. hydroelectric facilities represented 9 bgd (see P. Torcellini et al., *Consumptive Water Use for U.S. Power Production* (NREL, 2003), available at http://www.nrel.gov/docs /fy04osti /35190.pdf., hereafter referred to as NREL 2003). This estimate is not shown in the figure above because these reservoirs are used for multiple purposes; therefore, evaporation at these reservoirs cannot be attributed to hydropower alone.

Figure 2. Energy's U.S. Freshwater Consumption, (in gallons per day, bgd).

To illustrate the role that energy's water demand plays at the state level, roughly 16% of the water diverted in Kansas in 2008 went to biofuels; another 16% was used for electric power generation, while roughly 9% was used for municipal purposes.[26] Fourth, in some regions, climate change is anticipated to decrease the quantity and reliability of water supplies. In particular, lower-altitude and drier areas are anticipated to become drier; some regions (e.g., the West) may experience more frequent or intense drought years punctuated by wet years.[27] Lower precipitation in a portion of the Great Plains, and reduction in temperature-vulnerable snowpack in the western near-coastal mountains, are some of the anticipated changes to water supply.[28] Therefore, energy's demand for more water use may coincide with a decreasing and less predictable water supply.

Projections are based on assumptions that are subject to change; in particular, management of energy demand and wide deployment of water-efficient energy options may significantly lower water use below projected levels. A DOE Energy Efficiency and Renewable Energy (EERE) study found that expanding the nation's electricity portfolio to 20% wind by 2030 would reduce water consumption by 1.2 bgd compared to expanding the current electricity mix. The saved water would be 41% from the Midwest/Great Plains, 29% from the West, 16% from the Southeast, and 14% from the Northeast.[29]

Alternatively, the projections in Figure 1 could be underestimates. Some factors that may augment energy's water consumption are not accounted for in these projections. More water could be consumed if carbon capture and sequestration is widely employed (see discussion below), or if more water-intense energy is produced. For example, significant expansion of electricity from evaporative-cooled concentrating solar power (CSP), or new hydropower reservoirs in areas with high evaporation rates, could drive up energy's water consumption. Similarly, a significant increase in oil shale development could increase energy's water use in regions with those deposits. In summary, projections based on extrapolations of current trends can illustrate one potential path for water consumption, but many factors, such as the role of changing technologies and policies, can result in actual water consumption varying significantly from projections.

Carbon Capture May Create a New Energy Water Demand

The majority of current U.S. electric generation is from fossil fuels. The electric generation mix in 2009 was 45% coal, 23% natural gas, 20% nuclear, and 7% hydroelectric, less than 4% non-hydroelectric renewable generation, and less than 2% other sources.[30] Carbon capture technologies in the fossil fuel industry may reduce carbon dioxide emission, but they come with investment and resource costs to manufacture and operate. Current carbon capture technologies consume energy and water.[31]

Water is used at two points in the carbon capture cycle: cooling water is required for capture and compression processes, and water generally is consumed for power plant cooling when generating power needed to perform CCS. A 2009 study by the National Energy Technology Laboratory found that by 2030, carbon capture and sequestration could increase water consumption for electric generation by anywhere from 0.9 bgd to 2.3 bgd, depending on the scenarios used for plant additions and for deploying carbon capture technologies.[32]

Hydroelectric Water Consumption Is Poorly Documented

Although most hydropower generation represents an in-stream water use, dams built to generate hydropower and for other purposes consume water by increasing evaporation above free-flowing river conditions. That is, more evaporation occurs at the reservoir behind the dam than at the river without the dam.

How much evaporation occurs at a reservoir with a hydroelectric generation depends on site, climate, and water conditions. CRS was unable to locate national data on existing or future projections for water consumption for hydroelectric generation.

The currently available water consumption data from the USGS do not include hydropower evaporation. The agency states: "Although the quantity of water evaporated in the actual generation of hydropower (consumptive use) is small, considerable depletion of the available water supply for hydroelectric power generation occurs as an indirect result of evaporation from reservoirs and repeated reuse of water within pumped-storage power facilities."[33]

As noted in Figure 2, a 2003 NREL report estimated evaporation at 120 reservoirs with hydroelectric facilities at 9 bgd.[34] Because these reservoirs generally serve multiple purposes, this evaporation cannot be attributed to hydropower alone and is not shown in the figure.

Moreover, many reservoirs enhance water availability for multiple functions by providing storage that regulates streamflow, at times minimizing the occurrence of naturally occurring low-flow levels.

These benefits, however, often come with harm to aquatic ecosystems, species, and floodplains. The NREL data illustrate that evaporation at facilities with hydropower can vary widely based on geography and climate.[35] Site-specific estimates of water consumption for new hydroelectric generation using new reservoirs (e.g., some pumped storage proposals), therefore, would be required to understand hydroelectric generation's impacts on water resources. That is, new hydropower reservoirs cannot be assumed to have minimal effect on water consumption.

Efficiency improvements or additions of hydropower generation at existing facilities, however, have the potential to increase electricity generation without increasing water consumption from evaporation.

REGIONAL SIGNIFICANCE OF ENERGY'S WATER DEMAND

Much of the anticipated growth in the energy sector's water demand is in water-constrained areas, potentially exacerbating competition and low flow condition for water during summer and droughts. That is, while energy's water demand is anticipated to rise across the United States, the West is likely to experience some of the more significant constraints and conflicts in meeting this demand. While local or regional competition for water is often what makes energy's water demand significant, the regional and local scales of water resources and how they are managed often complicate federal water-related actions.

To illustrate one aspect of how energy's growing water demand varies across the country, Figure 3 shows projections for expanded electricity generation in many water-constrained states (e.g., California, Texas, Arizona), primarily owing to rising electricity demand.

The following examples of regional water consumption concerns are discussed in more detail: shale gas production in Texas using hydraulic fracturing, solar electricity generation in the Southwest; and biofuel production in the High Plains.

Technologies and management practices exist for reducing water use in each of these energy activities; however, these options generally come with energy cost, generation, and reliability penalties.

Whether the benefits from the water savings of adopting these technologies and practices are viewed as outweighing these penalties often is a matter of perspective and is site specific. Moreover, little reliable and systematic data is available on the costs and benefits of adopting these measures in a variety of circumstances to better inform decision-making.

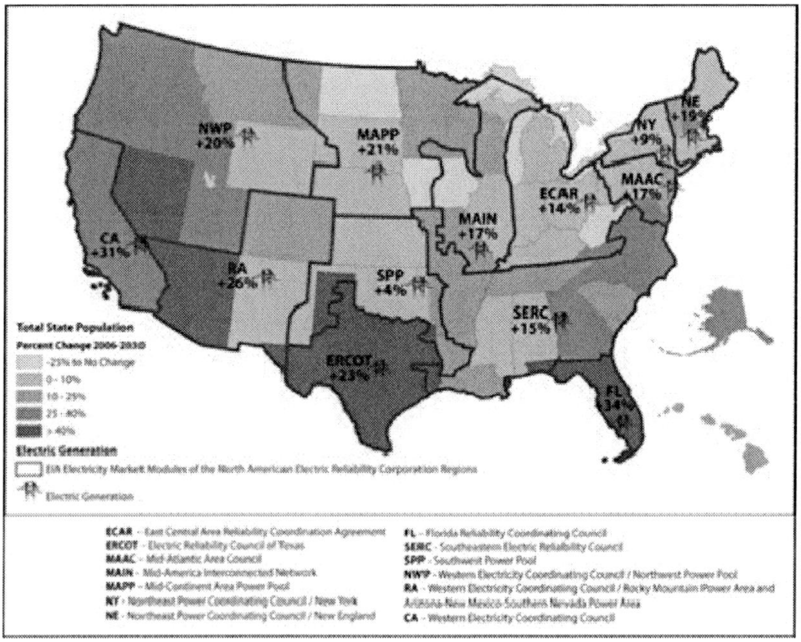

Source: EIA, *Supplemental Tables, Updated Annual Energy Outlook 2009 Reference Case with ARRA*, Wash., DC, April 2009, http://www.eia.doe.gov/oiaf/aeo/supplement/stimulus/suparra.htm); and U.S. Census Bureau, *U.S. Population Projections*, March 2004. EIA did not include Alaska and Hawaii.

Figure 3. Population and Electricity Generation Projections, 2010 to 2030.

Energy Development: Shale Gas in Texas

Shale gas development has expanded in the last decade, in part because of technological advances in horizontal drilling and hydraulic fracturing. Fracturing involves the pressurized injection of water-based fluids (water, sand, and chemical mixtures) into a well to fracture a rock formation so that natural gas is released. Shale gas formations occur in many areas of the United States.[36] Exploiting this resource is bringing the oil and gas industry into communities unaccustomed to energy development. Water use is one of a number of issues being raised with the expansion of unconventional gas development. While technological developments allow for greater extraction of gas, these technologies result in an increasing average amount of water used per well. That is, the longer the well, the more water is needed to drill and stimulate gas production. Freshwater, rather than saline water, is preferred for drilling and fracturing. Water use is concentrated in the early stages of well development, usually in the first few months. Once the well is producing, little or no water is required, unless refracturing is necessary. Much shale gas development is on private lands, and no government agency requires operators to report water use. Some data on water use per well are available, such as data from a DOE report in 2009 for four shale gas formations, which show water use ranging from 2.7 million gallons to 3.9 million gallons of freshwater per well.[37] Some limited data are available on the amount of water per unit of energy produced. Data from Chesapeake Energy shale gas wells indicate that drilling, fracturing, extraction, and processing use between 0.6 and 3.8 gallons of water per million British thermal units (MMBtu) produced, or 5 to 29 gallons per megawatt-hour (MWh), as fuel for a combined cycle natural gas facility (not including the power plant water use).[38] This puts shale gas at the lower end of water use for transportation fuels such as domestic onshore oil and irrigated biofuels, and below the water use for most coal and nuclear fuel production for electricity, according to a 2006 DOE report.[39]

The water use issue for natural gas is often one of concentration. That is, if many wells are being developed in a limited geographic area, the cumulative water needs of multiple drilling and fracturing operations may be locally significant and constraining to expansion of energy development. This may be particularly the case in areas with water constraints and competing water demands for domestic, agricultural, and thermoelectric use or areas where water for new gas operations represents a substantial expansion in water use.

Water quantity concerns have been raised for the Texas Barnett formation, particularly when drilling was using primarily fresh groundwater supplies and reached a peak of more than 3,000 new wells in 2008.[40] With the subsequent decline in gas prices, the number of new wells drilled annually has dropped to 1,000, and increasingly the water is coming from surface supplies purchased from municipal water utilities where available, such as in the Fort Worth area. Based on available projections, the maximum use for natural gas production annually in the affected Texas counties may reach 7.5 billion gallons (0.021 bgd) under a high drilling scenario; this would represent 1.15% of local water consumption.[41] While this shale gas use represents a modest share of local consumption, communities in this region of Texas have faced significant concerns about the sufficiency of available supplies during drought conditions for the region's growing population. This concern has brought scrutiny to all water uses, including shale gas. For most wells, the majority of the fluids injected into formation are not returned to the surface. In the Barnett formation, the water that is returned to the surface, known as flowback or produced water, is typically reinjected deep underground in a permitted disposal well. Texas A&M researcher C. J. Vavra estimates that more than half of the produced water could be reused in subsequent fracturing operations, and a quarter could be put to beneficial use.[42] These actions could reduce the impact of shale gas development on local water supplies but could raise other concerns. Legislation was considered by the 111th Congress to support research on this subject. For example, H.R. 469, the Produced Water Utilization Act of 2009, would have directed the Secretary of Energy to carry out a program to demonstrate technologies for environmentally sustainable use of energy-related produced waters from underground sources. Water quantity concerns related to shale gas development are often overshadowed by interest in the economic benefits of the gas development and other local concerns with development. These other concerns include the risk that fracturing-related activities may contaminate freshwater supplies, community disruption and air pollution from truck traffic related to gas development (e.g., truck transport of drilling equipment and materials and fracturing additives and freshwater on and wastewater off the well site), and short-term and long-term changes to the local landscape and community.

Although shale gas in Texas is discussed above, rising water demand for natural gas development is occurring or is anticipated in many other states and regions, such as the Northeast (Pennsylvania, New York), the South (e.g., Arkansas), and the upper Missouri River basin (Wyoming, South Dakota, North Dakota).

Renewable Energy Presents Regional Water Opportunities and Challenges

Water resource opportunities and challenges in meeting energy demand with renewable energy vary regionally.

- The Southwest has high-quality solar and geothermal resources, but water constraints and drought vulnerability favor more expensive, water-efficient use of these renewable resources.
- The Northwest generates most of its electricity from hydropower, but is diversifying to include more wind in light of environmental protections, increasing off-stream uses, and climate change effects on hydropower.

- The Great Plains may reduce water competition among the thermoelectric and agricultural sectors by
- exploiting wind power to produce electricity. Grid balancing and transmission of that electricity, especially to other regions, may pose constraints. Increases in irrigation to support bioenergy crops may tax stressed aquifers.
- The Southeast could reduce its dependence on coal and nuclear fuel and decrease its vulnerability to electricity interruptions during drought by producing electricity from biomass and photovoltaics.
- The Midwest could reduce competition between the thermoelectric and agricultural sectors and reduce the regional energy sector's low flow, drought, and flood vulnerability by exploiting its wind resources.
  This would require overcoming significant regulatory transmission issues. The Midwest is participating in the production of bioenergy crops, in particular corn for ethanol, raising water quality concerns, including that excess nutrients in agricultural runoff are feeding the growth of the "dead zone" in the Gulf of Mexico.
- The Northeast is not experiencing a regional issue with energy's growing water consumption; however, fracturing in shale gas formations is raising water quality concerns.
- Hawaii could transition to water-efficient power generation technologies to protect limited freshwater resources consumed in thermoelectric generation, which is dominated by oil-fueled power plants.
- Coastal regions, including Alaska, have the potential to develop offshore wind, tidal energy, waves, or ocean thermal gradients to reduce energy's onshore water use and land requirements. However, public opposition, transmission challenges, resource availability mismatched with demand, and natural hazards can pose challenges.

Source: CRS, with support from visiting researchers Ashlynn Stillwell and Kelly Twomey.

Electricity Generation: Solar in the Southwest

Another example of energy's regional water demand is concentrating solar power (CSP) in the Southwest. [43] The region has abundant solar resources, but the region's water constraints can influence the attractiveness and feasibility of different solar technologies and the suitability of specific sites. Most concentrating solar power (CSP) consists of ground-based arrays of mirrors that concentrate the sun's heat, which is used in a thermoelectric process to generate electricity. The two primary technologies are solar troughs and solar towers. Although CSP does not emit greenhouse gases, its use of a thermoelectric process can raise water concerns, particularly if evaporative cooling is employed (that is, if water is evaporated to dissipate waste heat). Similar water concerns would be raised if new fossil-fuel thermoelectric power plants were to be similarly located in the Southwest. For some Southwest counties with relatively low water use, large-scale deployment of CSP or another thermoelectric facility (even with water-efficient cooling technologies) could significantly increase the current demand for water in the county.[44]

Some solar developers are using cooling alternatives that require less water (or have been encouraged or required to do so as part of state permitting or federal approval of facilities on federal lands). These alternatives include dry or hybrid cooling or use of impaired waters for cooling (see Appendix C). These options generally come at a cost premium and with energy and cooling efficacy tradeoffs. Other solar developers have purchased water rights from willing sellers in states with active water markets. Still other solar developers are using solar technologies that require little or no water; these include photovoltaic solar, which uses panels of solar cells to convert sunlight directly into electricity, and dish engine CSP, which uses engines rather than a thermoelectric process[45] to produce electricity. (See Appendix C.) While these technologies are water-efficient, they have other constraints (e.g., cost, land use, dispatchability). In summary, freshwater constraints like those in the Southwest do not preclude solar development, but access to water shapes the technologies and costs of solar development.

Energy Crops: Biofuels in the High Plains[46]

An additional example of energy's regional water demand is biofuels in the High Plains. The expansion of biofuels use through incentives and consumption mandates is an example of how federal energy policies can affect national and local water use. The water quantity (and quality) used for biofuels is particularly sensitive to biofuel feedstock, use of irrigation, and local climate and soil conditions at the growing site. Average water consumption by the dominant U.S. biofuel, corn-based ethanol, significantly exceeds the water intensity of other U.S. transportation fuels if the corn feedstock is irrigated. (See Appendix B.) Irrigation of only a small amount of biofuel feedstock in areas without sufficient rainfall to support feedstock growth without irrigation has the potential to substantially increase national water consumption for transportation fuels. The High Plains—consisting of portions of Texas, New Mexico, Colorado, Kansas, Nebraska, Wyoming, and South Dakota—is one example of these low-rainfall areas. Much of the High Plains has faced water use issues for decades, such as the declining level of portions of the Ogallala aquifer since the mid-1960s.[47] Expansion of biofuels in this area is an additional demand exacerbating already competing water uses. For example, in Colorado in 2008, 16 times the annual quantity of treated water supplied to Fort Collins water customers was used to produce biofuels in the state, with the majority used in eastern Colorado.[48] Of particular concern to the High Plains is expanded biofuels production on new or marginal lands, which could lead to additional irrigation demand and increased nutrient application, causing both water quantity and quality concerns.[49] Even expansion of existing biofuel crops in current production, such as corn, raises water quality concerns with the possible increased application of fertilizers and pesticides necessary to increase yields (this is also of concern in regions such as the Midwest, where irrigation water concerns have been less significant).[50]

Recognition of water, land, and other issues related to biofuels, particularly irrigated corn- and soybean-based biofuels, has led to a search for feedstocks and other organisms that use fewer resources to produce.[51] Recent federal biofuels policies have attempted to assist this search by focusing on the development of a cellulosic biofuels industry.[52] Dedicated biomass crops, such as switchgrass, hybrid poplars, and hybrid willows are considered by many to be more desirable crops because they have a short rotation (regrow quickly after each harvest) and use fewer resources, such as water and fertilizers, than traditional field crop production. Despite potential environmental benefits, concerns persist about the additional use of fertilizers and water resources that could be required to increase the per-acre yields to economically feasible levels; for example, that cellulosic feedstocks may be irrigated to increase yields, even though irrigation may not be required.[53] Also, land use pressure for expanded production also applies to cellulosic biomass feedstock, possibly creating direct competition with current land conservation programs and replicating the concerns of traditional biofuel feedstock stated above. Despite federal incentives, technological and economic hurdles continue to prevent the cellulosic biofuels industry from developing to commercial scale production.[54]

MEETING AND MANAGING ENERGY'S WATER DEMAND: POLICY OPTIONS

The previous sections described how energy's demand for water is increasing and provided some regional examples. This section discusses options for meeting and managing that demand. Historically the energy sector and the states have determined how water is used in the energy sector, but the significant role that current federal policies are playing in driving up energy's water demand is raising questions about the federal role in meeting and managing that demand. These questions include:

- Are states being unfairly burdened with the responsibility of increased water use and competition resulting from federal energy policies, or is this part of the responsibility that comes with state primacy in water allocation?
- Who is responsible for the vulnerability of the nation's energy system to water availability?

Congress is faced with deciding not only whether, and if so how, to alter current policies to respond to energy's water demand, but also who is the most appropriate entity to respond to energy's growing water demand. Currently little direct federal action is aimed at managing the energy sector's water demand, although federal policies at times have significant indirect influence on this demand. Instead, present roles rely on the energy industry and the states and local governments to manage water constraints and to resolve energy-water conflicts.

The issue of the energy sector's water use may arise during the 112th Congress in a variety of contexts. Support or opposition for legislation affecting energy's water demand may be influenced by opinions about the proper federal role in water allocation and planning, as well as concerns about the cost of actions and who is responsible for those costs. Positions on the larger energy and climate debate and other factors may also be important. Federal responses to energy's water use are complicated by the wide-ranging and place-based nature of the issue, the variety of actors involved, the costs and other tradeoffs involved, and the existing institutions and divisions of responsibilities for water and energy.

If increased federal action to meet and manage energy's water demand is deemed appropriate, possible actions fall under a few broad options. Attempts can be made to minimize the growth in energy's freshwater use by adopting general energy policies that are less water intense and more sensitive to water constraints, or by specifically promoting activities that reduce energy's water use, such as incentives for adopting less water intense energy generation technologies. Another option is to make freshwater available for the energy sector (e.g., through allocations, permits, or facilitating water trading); however, the majority of water quantity allocation and permitting decisions are up to the states. An additional option is improving data and analysis on energy's water use to better inform decision-making (e.g., resource planning efforts and decision-support tools) and enhancing the availability and dissemination of water-efficient technological alternatives. These approaches are presented in Table 2, with examples of each from legislation reported by a congressional committee of the 111th Congress.

These options are not mutually exclusive and different options may be more or less appropriate and attractive for different components of the energy sector's water demand. These options also represent different potential roles and costs for federal and state governments, the energy sector, and energy consumers. No entity has performed a comparative analysis of these policy options using multiple criteria (e.g., cost-effectiveness; who bears the cost; risks, reliability, and vulnerability; opportunity costs; role of state entities; role of federal entities). One of the challenges of a federal response to energy's water demand is that the concerns, policy options, and technological options vary greatly by region.[55]

Table 2. Sample Legislative Responses to Energy's Water Demand

Response	Examples in Committee-Reported Legislation in the 111th Congress
No Federal Action Directed at Managing Water Use	Existing federal energy (including conservation and efficiency policies), climate, agricultural, and environmental laws and policies indirectly, but at times significantly, shape the energy sector's water use.
Minimize Energy Sector's Growth in Water Use	

Response	Examples in Committee-Reported Legislation in the 111th Congress
Promote water-efficient energy through standards, regulations, or incentives	Could be an outcome related to increased adoption of electric generation from wind, photovoltaics, or natural gas—which may occur as a result of energy markets, cap and trade legislation (e.g., H.R. 2454, American Clean Energy and Security Act of 2009, passed House and placed on Senate Legislative Calendar), or renewable electricity standards (e.g., S. 1462, American Clean Energy Leadership Act of 2009, reported by Senate ENR)— depending on how the energy industry responds to the legislation.
Promote use of technologies to reduce energy's water use though incentives or regulations	No explicit mention in legislation reported in the 111th Congress; however, state and federal energy permitting actions have required or promoted actions to reduce water use (e.g., roughly half of fast-tracked CSP development projects on federal lands are adopting dry cooling because of concerns about freshwater demands of other cooling options).
Promote conservation and efficiency measures through incentives or regulations	§ 2 of S. 3396, Supply Star Act of 2010 (reported by Senate ENR), would establish a federal program to promote resource-efficient industrial supply chains (such sums as necessary, $35 million total for five-year CBO authorization estimate).
	§ 147 of Title I, Subtitle D, "Energy and Water Integration," in S. 1462 (reported by Senate ENR) would direct the Secretary of Energy to create a competitive energy-water conservation grant program for local governments, water agencies, and tribes ($100 million for each fiscal year from FY2010 to FY2015).
Develop and assess technologies to reduce energy's water use	§ 142 of S. 1462 would require a federal study to identify alternatives to optimize water and energy efficiency in the production of electricity (such sums as necessary). § 162 of S. 1733, Clean Energy Jobs and American Power Act (reported by Senate EPW), would require EPA to create a "1,000,000,000 Gallon Challenge Grant Program" to fund projects that have the potential for producing significant quantities of biofuels from renewable biomass, where surface and groundwater use and impacts are considered in determining whether the biofuel is renewable ($500 million for FY2010 through FY2014).

Table 2. (Continued)

Response	Examples in Committee-Reported Legislation in the 111th Congress
	§ 196 of Title I of H.R. 2454 and § 152 of S. 1733 would authorize the Secretary of Energy to provide nonprofit entities with grants to conduct competitive programs to support start-up businesses in clean energy, including water conservation ($20 million).
Improve Energy Sector Access to Water	
Allocate sustainably available water	Largely state responsibility; no explicit mention in federal legislation, but some federal actions underway (e.g., review of requests for surplus water contracts under Flood Control Act of 1944, 33 U.S.C. § 708). Also, already authorized efforts, if completed, could inform allocation decisions; for example, § 9503 of Subtitle IV, "Secure Water," in P.L. 111-11 authorizes the assessment and development of strategies to address water quantity effects of global climate change in Bureau of Reclamation service areas.
Facilitate transfer of water saved in non-energy sector	§ 9504 of P.L. 111-11 establishes a program to provide grants or enter into cooperative agreements to (among other purposes) conserve water, increase water use efficiency, and facilitate water markets in the 17 Reclamation states ($200 million).
Inform Decisions and Support Research Expanding Water-Efficient Energy Options	
Data and Assessments	§ 9508 of P.L. 111-11 requires the Secretary of the Interior to establish a national water availability and use assessment program that reports on significant trends, including significant changes in water use due to the development of new energy supplies ($20.0 million for each of FY2009 through FY2013). § 3 of H.R. 3598, Energy and Water Integration Act (passed House), would create a council to recommend actions that promote improved energy and water resource data collection, and to conduct workshops to promote data exchange ($5 million for each of fiscal years 2011 through 2015).

Response	Examples in Committee-Reported Legislation in the 111th Congress
	§ 149 of S. 1462 would mandate a study, led by the Department of Energy, examining industrial water use, peak energy use in water treatment and delivery, and energy "embedded" in water (no authorization of appropriations specified). § 9505 of P.L. 111-11, requires the Secretary of Energy to assess federal hydroelectric power production to water supply risks posed by global climate change (such sums as necessary).
Research	§ 2 of H.R. 3598 would require DOE to develop a strategic plan for energy-water research needs within DOE research activities ($60 million for each of FY2011 through FY2015). § 171 of Subtitle H, Title I, of H.R. 2454 would direct the Secretary of Energy to establish energy innovation hubs to conduct research, including research that enhances water security (no specific sum set).
Integrated energy - water planning	No explicit mention in legislation in the 111th Congress
Technical assistance	§ 148 of S. 1462 would require the Secretary of Energy to create a technical assistance program for reducing energy use by rural water utilities ($7 million for each of FY2010 through FY2015). § 10 of H.R. 1145, National Water Research and Development Initiative Act of 2009, would require the EPA Administrator to establish a pilot program for energy audits of water related infrastructure (no amount specified for this section, entire bill has $2 million for each of FY2010 through Fy2014).

Source: CRS.

Notes: ENR = Energy and Natural Resources Committee, EPW = Environment and Public Works Committee, EPA = U.S. Environmental Protection Agency.

Whether these policy alternatives should be pursued at the local, state, or national level depends in part on perspectives on the appropriate role of each level of government. Perspectives on the policy options related to energy's water demand also are influenced by long histories of regulation, management, promotion, and oversight of the nation's energy and water resources and infrastructure. Current and evolving conditions also play a role: providing greater access to water for the energy sector (which is primarily up to the states) may be difficult and controversial in water-scarce, drought-prone, or environmentally sensitive areas, especially with climate change anticipated to affect the availability and reliability of water in some regions.

States have traditionally had primacy in water allocation, so decisions about permitting energy's water use largely have been non-federal. State and federal laws and policies can affect the ease or difficulty of, and incentives for, transferring water from existing uses to energy.

The federal government can, if it chooses, promote change in state water laws, institutions, and decisionmaking.[56] Some view this as infringing on states' rights.[57] Similarly, while private entities make many of the decisions on the energy sector's water use, the public sector influences these decisions through numerous routes (e.g., tax incentives, loan guarantees, permits, regulations, planning, education); this influence can come from local, state, or federal policies.

The federal role in water resource allocation and management increases as the federal interest increases—for example, when the use occurs on federal lands. Competing water demands, including those from the energy sector, are raising questions for federal agencies about the operations of federal facilities. For instance, can water delivered to Bureau of Reclamation contractors (e.g., water delivered to irrigation districts in California's Central Valley) be used not for agriculture, but for energy development (e.g., evaporative cooling of a concentrating solar power facility)?

Can water from an Army Corps of Engineers dam that has multiple purposes (e.g., flood control, navigation, and/or municipal or agricultural water supply) be used for the oil and gas development, and if so, how much water can the Corps provide under existing authorities?

Pursuit of Water Stored Behind Federal Dams for Oil and Gas Water Supply

Western North Dakota illustrates some stakeholders' interest in using federal infrastructure to satisfy the energy sector's water demand. The region has experienced substantial expansion in oil and gas production since 2008.

The industry's largely new water demand is anticipated to soon reach 24 million gallons per day (27,000 acre-feet annually). At that point, it would represent one-third of total water usage in the 11 counties in Western North Dakota that are producing oil and gas and located in proximity to a lake formed by a federal dam on the Missouri River.

In December 2010, the Corps released a draft report, *Garrison Dam/Lake Sakakawea Project, North Dakota, Draft Surplus Water Report* (Dec. 2010), that if finalized would allow the Secretary of the Army to meet the industry's water demand for up to 10 years by allowing withdrawals from Lake Sakakawea at Garrison Dam (ND). While other options for accessing the water sought by the oil and gas industry are available, the draft report finds that the Corps' provision of water from the lake would be the least costly means of addressing the water demands of the oil and gas industry and other water users in the region.

Observations and Concluding Remarks

The energy sector has long been a major water user, so why the current concern? Major energy trends are pushing the energy sector to become more dependent on, and therefore vulnerable to, freshwater availability. This is occurring at a time of increasing concerns about the adequacy and reliability of freshwater supplies due to population growth and climate change. Energy resource and technology paths chosen and capital investments made in the near term are likely to establish long-term trajectories for energy's water use.

Many of the current trends are in part driven by federal policy. The federal government partially shapes the energy sector and at times defines a vision for the nation's energy future (e.g., biofuel production targets). Some stakeholders have raised concerns about the feasibility and consequences of meeting various energy targets and policy proposals, including concerns about physical inputs like water, land, materials, and rare minerals.[58] Because affordable freshwater is a finite resource, commitments of water supplies for the energy sector reduce availability for other sectors and for ecosystems. Local or regional competition for water is often what makes energy's water demand significant; at the same time, it is the regional and local scale of water resources and how they are managed that often complicates many federal water-related actions. The federal role in energy supply and demand raises questions about the policy direction for meeting and managing energy's water needs, among them: If energy security is a national security issue, is energy's water use by association a national security issue? Would this be a justification for federal spending on energy and water efficiency measures?

Water supply concerns are not only being raised in the context of energy. There also is growing concern about water availability and aquatic conditions for meeting the demands of the agricultural and municipal sectors and the needs of ecosystems and threatened and endangered species; these concerns are raised particularly in the context of droughts and impacts of climate change on water resources. Given available freshwater resources, a challenge for the nation will be to cost-effectively, sustainably, and reliably meet energy demands while satisfying agricultural, municipal, and industrial water demand, as well as ecosystem needs. This challenge raises fundamental and controversial questions about how U.S. freshwater resources are allocated and used for various purposes, and about the availability and value of water in different sectors of the economy and in the environment. At issue for Congress is the role that the federal government and federal funding plays in shaping, meeting, and managing energy's water demand, while accounting for the significant role of the states and private sector in water use decisions.

APPENDIX A. ENERGY'S WATER CONSUMPTION TRENDS

Ideally, policy and decision makers should know how energy choices and policies compare across a wide array of parameters (e.g., cost, reliability, dispatchability, emissions, land requirements, water use) under different scenarios. That is, informed decisions would require water use data, analyses of least-cost water and energy conservation and efficiency actions, understanding of other water uses and their costs and benefits, life-cycle assessments of energy's water uses and risks, and more. Such assessments are not available. While not addressing this shortfall, Table A-1 and Table A-2 summarize the freshwater impacts of numerous shifts in transportation fuels and the electricity sector, respectively.

Without such multi-variable assessments, it is difficult to understand the full implications of energy decisions. While this report focuses on energy's water use, there are a host of other factors to consider when analyzing energy policy tradeoffs. For example, fuel and technologies for generating electricity are not equally dispatchable; that is, is they are not equally able to increase or decrease generation, or to be brought online or shut down to match demand. Thermoelectric facilities using fossil fuel or geothermal sources are advantageous because they can be operated to produce electricity constantly or dialed up or down with demand. While electricity from hydropower, tidal, and wave energy generally can be produced fairly predictably, often the timing of operations is subject to the intensity of tides and waves, to water conditions such as drought or large storms, and to environmental restrictions regarding potential impacts on marine life and ecosystems.

In contrast, until more advanced storage technologies become commercially competitive, photovoltaics and wind, which require very little water, remain intermittent sources that generally cannot be dispatched when the sun is not shining or the wind is not blowing.

Table A-1. Domestic Freshwater Impacts from Transportation Fuel Shifts

Shifts in Imported Fuel Use	Change in Domestic Freshwater Consumption	Impairments of domestic water quality and/or aquatic ecosystems
Domestic fuel sources displace foreign fuels	Little change or ⇧ depending on domestic fuel	⇧
Domestic oil using enhanced oil recovery displaces foreign gasoline	⇧	⇧ from produced water
Domestic offshore oil displaces foreign oil	Little change	⇧ from spill risk
Fuel conservation (e.g., resulting from increased fuel efficiency standards) decreases oil imports	Little change	Little change
Imported biofuels displace foreign gasoline	Little change	Little change

Source: CRS.

Table A-2. Domestic Freshwater Impacts from Electricity Sector Shifts

Shifts in Electricity Generation	Change in U.S. Freshwater Consumption	Impairment of Domestic Water Quality and/or Aquatic Ecosystems
Fuel Source Shift		
Nuclear displaces coal	Unclear with current data	Shift in pollutants generated during fuel extraction, processing, and electricity generation
Concentrating solar power (with freshwater cooling) displaces coal	⇧	⇩

Table A-2. (Continued)

Shifts in Electricity Generation	Change in U.S. Freshwater Consumption	Impairment of Domestic Water Quality and/or Aquatic Ecosystems
Domestic natural gas displaces coal	↓	Shift from nonpoint source to more point source pollution
Hydropower without new reservoirs displaces coal	↓	Shift from nonpoint source pollution to potential harm to aquatic ecosystem and species
Photovoltaics displace coal	↓	↓
Wind displaces coal	↓	↓
Ocean/tidal/hydrokinetic displace coal	↓	Potential for harm to aquatic ecosystem and species
Geothermal displaces coal	↓	↓
Electricity Sector Shifts		
CCS added to a power plant	↑	unknown
Dry cooling replaces evaporative cooling	↓	↓
Electricity conservation decreases electricity consumption	↓	↓

Source: CRS.

APPENDIX B. WATER USE BY TRANSPORTATION FUELS

D. Elcock's study projects that the energy sector's consumptive water use will increase from 6 billion gallons per day (bgd) in 2005 to 10 bgd in 2030 in the areas of fossil fuel mining, production, and processing and bioenergy.[59] (See Figure B-1.) The growth is dominated by water consumed for bioenergy.[60] The Elcock study and other sources show that the effect of bioenergy on energy's water use was already felt between 2005 and 2010.[61]

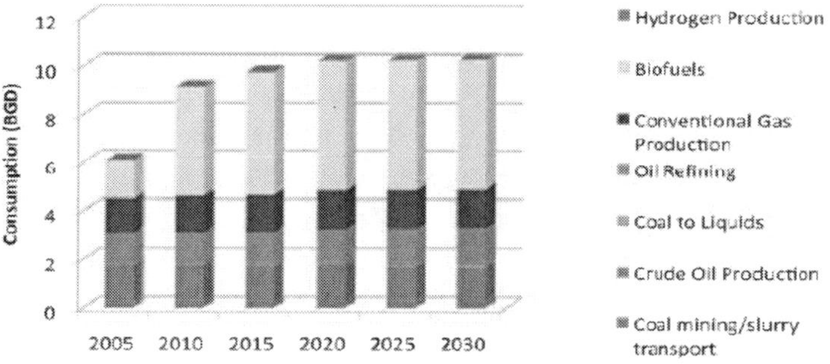

Source: Elcock 2010.

Figure B-1. Biofuels Dominate Energy Production's Projected Water Consumption.

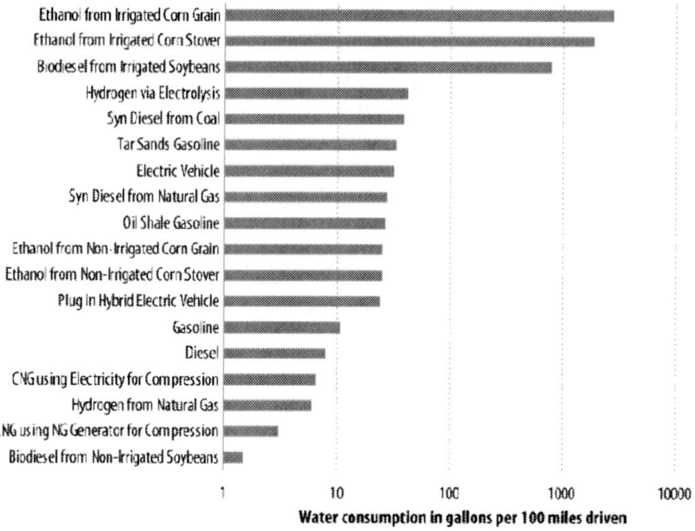

Source: CRS, modified from M. Mantell, "Abundant, Affordable, and Surprisingly Water Efficient," presented at 2009 GWCP Water/Energy Sustainability Symposium in Salt Lake City, September 13-16, 2009. Original data from C. King and M. Webber, "Water Intensity of Transportation," *Environmental Science & Technology*, vol . 42, no. 21 (2009), available at http://pubs.acs.org/doi/pdf /10.1021/es800367m.

Notes: NG = natural gas; CNG = compressed natural gas. Note the logarithmic scale.

Figure B-2.Water Intensity of Transportation Fuels.

The projected water demand by bioenergy could drop, potentially significantly, if less-water intense bioenergy is developed and adopted. This projection covers most of the water used to fuel transportation and the water used for obtaining and preparing fossil fuels for use in either transportation or the electricity sector. Figure B-2 provides data from research on the comparative water intensity of different U.S. transportation fuels. Note that the figure has a logarithmic scale, and that the water consumed traveling on irrigated biofuels is orders of magnitude larger than the water embedded in other fuels.

APPENDIX C. WATER USE FOR ELECTRICITY GENERATION

Trends in the electricity sector contributing to increased water consumption currently overwhelm actions that reduce water consumption. As a result, local water challenges presented by electric generation are becoming more common in many water-constrained areas. These challenges may drive action to reduce electricity's freshwater footprint, such as the adoption of water-efficient electricity generation and water-efficient thermoelectric cooling options, or the use of impaired water. This appendix describes the available data on the electricity sector's water consumption and data on the different electricity generation's water use. Then it presents alternative thermoelectric cooling choices that may reduce electricity's water use. Lastly, it discusses the tradeoffs between water use and other characteristics (e.g., environmental impacts, reliability, dispatchability) of electric generation from hydropower facilities, photovoltaic solar and wind installations, and geothermal resources.

Data on Electricity's Water Use

Available estimates for the electricity sector's water consumption focus primarily on thermoelectric cooling water needs because cooling dominates water use during generation. Other aspects of electric production (e.g., fuel mining and processing) also use water. (See Table C-1.) Data and projections on the water consumed in the mining, production, and processing of fuels for generating electricity are bundled with the production of fossil fuels used in the transportation sector, like the Elcock study. The projections in Figure 2 for thermoelectric cooling come from the Elcock study, which was based on projections in a 2007 report by NETL.[62] The NETL projections are limited to coal, natural gas, and nuclear-powered facilities; that is, they do not calculate water use changes that would occur from shifts in the electricity sector broadly (e.g., adoption of more wind or photovoltaics) or shifts in thermoelectric generation more specifically (e.g., concentrating solar power or biomass generation). Elcock assumed that biomass generation for electricity production would be based in regions not requiring irrigation.[63]

No analyses of electricity's total water consumption or how it may change under different policies exist; this type of analysis is difficult to perform without reliable data on water intensities of different electricity generation technologies and fuels. No authoritative comparison of the various water intensities of electricity generation options exists. Figure C-1 and Figure C-2 illustrate the best available data for average water intensities for various electricity alternatives, with and without carbon mitigation using CCS, respectively. These figures are based on Table C-1, which represents the best available data on water consumed in producing electricity using various fuels and technologies.[64] The data in Figure C-1 are imperfect and come from multiple sources, raising questions about whether the data can be used to make accurate comparisons across technologies.

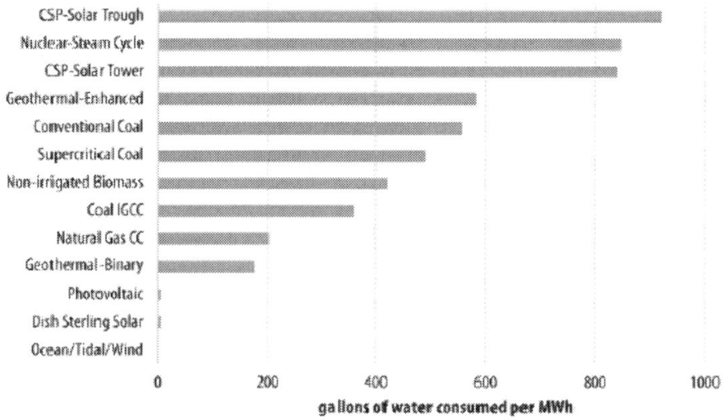

Source: CRS. Data represent evaporative-cooled thermoelectric, except for geothermal, see Table C-1. IGCC = Integrated Gasification Combined-Cycle; CC = Combined Cycle.

Figure C-1.Water Intensity of Electricity Generation Alternatives without CCS.

Table C-1. Water Consumption for Electricity Generation by Fuel Source

Generation Technology and Fuel	Water for Fuel Mining, Production, or Processing (gal/MWh)	Evaporative Cooling Water at Power Plant (gal/MWh)	Other Water Used for Power Plant Operations (gal/MWh)	Avg. Total Water Intensity (gal/MWh)	Data Sources
Biomass/Waste					
Non-irrigated biomass	0	300-480	30	420	DOE 2006
Non-irrigated biomass with CCS	0	Not available	30	—	DOE 2006
Irrigated biomass	Highly variable	300-480	30	—	DOE 2006
Municipal waste	Not available	300-480	30	420	DOE 2006
Coal					
Conventional/Subcritical coal	5-74	449	68	557	DOE 2006; NETL 2009
Conventional/Subcritical coal with CCS	5-74	884	101	—	DOE 2006; NETL 2009
Supercritical coal	5-74	392	59	491	DOE 2006; NETL 2009
Supercritical coal with CCS	5-74	759	86	—	DOE 2006; NETL 2009
Ultra-supercritical Coal Ultra clean coal	Not available	Not available	Not available		
	Not available	Not available	Not available		
Coal IGCC (slurry fed)	30-70	290	19	359	DOE 2006; NETL 2009
Coal IGCC (slurry fed) with CCS	30-70	355	97	502	DOE 2006; NETL 2009

Generation Technology and Fuel	Water for Fuel Mining, Production, or Processing (gal/MWh)	Evaporative Cooling Water at Power Plant (gal/MWh)	Other Water Used for Power Plant Operations (gal/MWh)	Avg. Total Water Intensity (gal/MWh)	Data Sources
Coal IGCC (dry fed)	5-74	243	53	336	DOE 2006; NETL 2009
Coal IGCC (dry fed) with CCS	5-74	355	120	515	DOE 2006; NETL 2009
Geothermal					
Enhanced geothermal	0	Dry cooling	290-720	585	Argonne 2010
Geothermal binary	0	Dry cooling	80-270	175	Argonne 2010
Geothermal flash	0	Dry cooling	5-10	8	Argonne 2010
Hydroelectric			Evaporation varies with site		NREL 2003
Natural Gas					
Natural gas combined-cycle Natural gas combined-cycle with CCS	11	192 338	0 0	203	DOE 2006; NETL 2009
	11			349	DOE 2006; NETL 2009
Nuclear					
Nuclear steam cycle	45-150	720 (cooling pond)	30	848	DOE 2006
Nuclear boiling water	45-150	Not available	—	—	DOE 2006
High temperature gas cooled nuclear	45-15	Not Available	—	—	DOE 2006
Small modular nuclear	Not Available	Dry cooling possible	Not available	—	
Ocean/Tidal	0	Not applicable	Not available	0	

Table C-1. (Continued)

Generation Technology and Fuel	Water for Fuel Mining, Production, or Processing (gal/MWh)	Evaporative Cooling Water at Power Plant (gal/MWh)	Other Water Used for Power Plant Operations (gal/MWh)	Avg. Total Water Intensity (gal/MWh)	Data Sources
Oil					
Oil	No data	300-480	Not available	—	DOE 2006
Oil from enhanced oil recovery	Highly variable, potentially significant	300-480	Not available	—	DOE 2006
Solar					
CSP – solar trough	0	760-920	80	920	DOE 2006; DOE n.d.
CSP – solar tower	0	750	90	840	DOE 2006; DOE n.d.
Dish Engine Solar	0	Air cooled	4	4	NREL 2002
Photovoltaics	0	Not applicable	4, higher for locations requiring more panel washing	4	NREL 2002
Wind – onshore or offshore	negligible		0		DOE 2008

Sources: Argonne National Laboratory released for comment the following report: ANL, *DRAFT Water Use in the Development and Operation of Geothermal Power Plants*, Sep. 2010. *Energy Demands on Water Resources: Report to Congress on the Interdependency of Energy and Water*, Dec. 2006. DOE, *Concentrating Solar Power Commercial Application Study: Reducing Water Consumption of Concentrating Solar Power Electricity Generation*, Washington, DC, no date (n.d.). NETL, *Estimating Freshwater Needs to Meet Future Thermoelectric Generation Requirements (2009 Update)*, 2009. NREL, *Fuel from the Sky: Solar Power's Potential for Western Energy Supply*, NREL/SR-550-32160 July 2002. P. Torcellini et al., *Consumptive Water Use for U.S. Power Production* (NREL), 2003.

Notes: Evaporative cooling for thermoelectric facilities is assumed, unless otherwise noted in the table. Data do not include water consumed in manufacturing or construction. While the total column represents the average total water consumption for a unit of electricity, the water for fuel production may occur at a different location than the water consumed at the power plant.

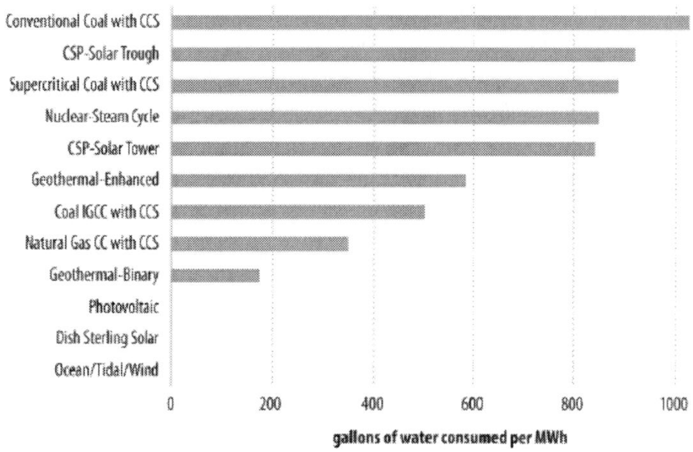

Source: CRS. Data represent evaporative-cooled thermoelectric, except for geothermal, see Table C-1.

Figure C-2.Water Intensity of Electricity Generation Alternatives with CCS.

Thermoelectric Cooling: Emerging Alternatives

The withdrawal and water quality impacts of once-through cooling have resulted in newer power plants generally using evaporative cooling.[65] Emerging cooling technologies have the potential to use much less freshwater than once-through or evaporative cooling. These include dry cooling (previously discussed), hybrid dry-wet cooling, cooling with fluids other than freshwater, and more innovative technologies (e.g., wet-surface air coolers, advanced wet cooling). However, these alternatives have their own costs and disadvantages.

A DOE report found that dry cooling could reduce water consumption to roughly 80 gal/MWh for solar troughs and 90 gal/MWh for solar towers.[66] While they consume less water, dry and hybrid cooling systems have financial as well as efficiency costs. Total annualized costs for dry cooling tower systems can be four times those of evaporative cooling tower systems.[67] Due to the higher cooling and lower generation efficiency costs, the cost of electricity from a dry cooled plant may be 10% higher than a similar wet cooled plant.[68] Dry cooling uses fans to blow air for steam condensation. While power plants with dry cooling use considerably less water, dry cooling is less effective at cooling the power plant than evaporative cooling, thus reducing electric generation at the facility. The DOE report also found that electricity generation at a dry-cooled facility dropped off at ambient temperatures above 100°F. For a solar parabolic trough facility in the Southwest, the benefit in the reduction in water consumption from dry cooling resulted in cost increases of 2% to 9% and a reduction in energy generation of 4.5% to 5%.[69] The cost and energy generation penalties for dry cooling depend largely on how much time a facility has ambient temperature above 100°F. Dry cooling would reduce generation on the same hot days when summer peak electricity demand is greatest.

Hybrid wet-dry cooling attempts to balance water consumption with power generation efficiency; it remains under development for commercial scale applications. To weigh the tradeoffs in energy generation, cost, and water use, DOE researched hybrid cooling processes that combine dry and evaporative cooling. The hybrid system consists of parallel evaporative and dry cooling facilities, with the evaporative cooling operating only on hot days. By using dry cooling generally and evaporative cooling above certain ambient temperatures, losses of thermal efficiency from dry cooling can be reduced. How often the evaporative cooling is used determines how much water is consumed and the effect of hot days on thermal efficiency. DOE found that a hybrid cooling system in the Southwest using 50% of the water of evaporative cooling would maintain 99% of the performance of an evaporative-cooled facility. A hybrid cooling system using 10% of the water of evaporative cooling would maintain 97% of the energy performance.

Another means for decreasing freshwater impacts is employing alternative water sources for evaporative-cooling.[70] These alternative water sources include effluent from wastewater treatment plants[71] or other reclaimed or impaired water, such as brackish or low-quality groundwater[72] and mine pool water.[73] These alternative water sources, however, may lead to additional scaling, corrosion, and fouling of cooling equipment or require pretreatment before cooling use.

Additional research may be able to improve the viability of saline water cooling. Availability of alternative water sources in proximity to electricity generation facilities is a potential limitation to their use. Understanding of the opportunities for brackish cooling alternatives is likely to be improved when the assessment of brackish groundwater authorized by Section 9507 of P.L. 111- 11, the Omnibus Public Land Management Act of 2009, becomes available.

Tradeoffs of Select Electricity Generation Technologies

Hydroelectric Generation

Hydroelectric power is produced when water passes through a turbine. Turbines for large-scale hydroelectric generation are located at dams. Electricity at U.S. hydropower facilities is produced with relatively low greenhouse gas emissions. However, hydropower's environmental effects can be significant. Conventional hydropower development through dam building often significantly alters river ecosystems, harming many indigenous species. Drought and changes to hydrology, such as possible reduction in snowpack under a changing climate, can reduce electricity generation at hydropower facilities because of the effects on reservoir operations and levels.

Constructing new large dams is contentious; therefore, efforts to identify opportunities for increasing hydropower generation have focused on smaller-scale opportunities or improved efficiency and expansion of hydropower at existing facilities.[74] The Electric Power Research Institute estimates a potential capacity hydropower gain of 10 GW by 2025 as feasible, without the construction of new large dams.[75] Six western states—Alaska, Washington, Oregon, California, Idaho, and Montana—have the highest potentials.[76] Despite this identified potential, little additional hydropower generation has been installed in recent years for a number of reasons (e.g., hydropower and water-related permit and regulatory requirements, and public concerns and perceptions about environmental effects). Similarly, the number of applications for FERC preliminary permits for pumped storage has increased substantially in recent years; however, the proposals have not proceeded to construction.

Photovoltaic Solar and Wind

Renewable electricity technologies that do not use thermoelectric processes may have minimal water requirements for electricity generation. Wind turbines and solar photovoltaic (PV) panels, for example, require small volumes of water for cleaning, but otherwise use no water. However, the minimal water intensity of wind and PV comes with tradeoffs. Transmission constraints;[77] cost; and regulatory, technical, and operational factors currently restrict the extent to which solar and wind resources can be exploited to meet electricity demand. As previously noted, wind and PV are intermittent electricity sources. Some storage options for these intermittent technologies exist (e.g., wind used in conjunction with a pumped storage hydropower facility); however, their applications are limited and intermittency continues to limit generation from wind and PV.[78] Electricity from PV is also currently more expensive than electricity from CSP, although electricity from wind is less expensive than electricity from CSP.

Geothermal

There are several ways to use geothermal energy: electricity generation (discussed herein) as well as direct-use (recovering water heated by the earth) and heat pumps (using the earth's heat to cool/heat buildings). Traditional geothermal power production uses naturally occurring convective hydrothermal sources in hot rock formations to produce steam to run a thermoelectric power plant's turbines (i.e., a geothermal flash system). Alternatively, for lower temperature geothermal resources, a second working fluid is heated by the geothermal water using a heat exchanger; it is the working fluid that drives the turbines (i.e., a geothermal binary system). Finally, because the majority of hot rock is dry, electricity can also be generated by injecting water into fractured rock to be heated (i.e., an enhanced geothermal system).[79]

The water is then injected back into the rock formation to create a relatively closed-loop system. Because the geothermal or injected water is an essential component of geothermal electricity generation and the size of the plants are generally smaller than 50 MW, dry cooling is becoming the standard for new geothermal facilities. Smaller power plants are generally easier to dry-cool than larger plants.

A 2008 USGS study estimated that the United States has geothermal resources sufficient to supply half of the nation's electric generation needs, assuming enhanced geothermal systems could be successfully developed and deployed at a commercial scale.[80]

That is, much of the geothermal potential shown on maps of geothermal resources[81] requires water to be injected (and therefore consumed) to exploit the geothermal energy. Enhanced geothermal power plants require relatively little land and can be used in coproduction with enhanced oil recovery to lengthen the lifespan of oil fields.[82] These enhanced systems are an emerging technology, so more research and development are needed for large-scale commercial deployment.

Current research is investigating the possibility of replacing water with carbon dioxide as the working fluid, which would significantly reduce water usage and would be a means of carbon sequestration.

End Notes

[1] Thermoelectric power plants burn or react fuel to generate steam, which turns a turbine connected to a generator that produces electricity. Cooling water is used to condense the steam into boiler feed water, so the process can be repeated. See Appendix **C** for more information on thermoelectric cooling technologies and water use.

[2] U.S. Geological Survey, *Estimated Use of Water in the United States in 2005* (Circular 1344: 2009).

[3] D. Elcock, "Future U.S. Water Consumption: The Role of Energy Production," *Journal of the American Water Resources Association*, vol. 46, no. 3 (June 2010), pp. 447-480, hereafter referred to as Elcock 2010.

[4] See CRS Report RS21421, *Mountaintop Mining: Background on Current Controversies*, by Claudia Copeland

[5] See CRS Report R41311, *The Deepwater Horizon Oil Spill: Coastal Wetland and Wildlife Impacts and Response*, by M. Lynne Corn and Claudia Copeland.

[6] DOE, National Energy Technology Laboratory (NETL), *Estimating Freshwater Needs to Meet Future Thermoelectric Generation Requirements (2009 Update)*, p. 64, http://www.netl. doe.gov/energy-analyses/pubs/ 2009%20Water%20Needs%20Analysis%20-%20Final%20 (9-30-2009).pdf, hereafter referred to as NETL 2009.

[7] EIA, *Annual Energy Outlook 2010 Early Release*, Washington, DC, http://www.eia.doe.gov /oiaf/aeo/aeoref_tab.html.

[8] For more on oil shale and water rights, see CRS Report RS22986, *Water Rights Related to Oil Shale Development in the Upper Colorado River Basin*, by Cynthia Brougher.

[9] U.S. Department of Agriculture (USDA) Forest Service, *2000 RPA Assessment of Forest and Range Lands*, FS-687, February 2001, p. 14.

[10] See U.S. Climate Change Science Program and the Subcommittee on Global Change Research, *Effects of Climate Change on Energy Production and Use in the United States*, Synthesis and Assessment Product 4.5, February 2008, pp. 31-32, hereafter referred to as U.S. Climate Science Program 2008.

[11] M. Hightower and S.A. Pierce, "The Energy Challenge," *Nature*, vol. 452, no. 7185 (March 20, 2008), pp. 285-286.

[12] Tennessee Valley Authority, "President's Report" presentation, Nov. 2010, p. 88, http://www.tva.gov/abouttva/ board/pdf/11-4-2010_board_final.pdf.

[13] Ibid, p. 89.

[14] U.S. Climate Science Program 2008, p. 45, found that "hydroelectric power generation can be expected to be directly and significantly affected by climate change...with potential for production decreases in key areas such as the Columbia River Basin and Northern California." Hydroelectric generation also may decrease if dams are reoperated to maximize stored water and minimize losses to evaporation, in areas where conserving available freshwater supplies may become more critical.

[15] M. Furniss et al., *Water, Climate Change and Forests: Watershed Stewardship for a Changing Climate*, U.S. Department of Agriculture, Forest Service, Pacific Northwest Research Station, PNW-GTR-812, Portland, OR, June 2010, http://www.fs.fed.us/pnw/pubs /pnw_gtr812.pdf, hereafter referred to as Forest Service 2010.

[16] U.S. Climate Science Program 2008, p. 38.

[17] For a discussion of data gaps related to thermoelectric power plants, see U.S. Government Accountability Office (GAO), *Energy-Water Nexus: Improvements to Federal Water Use*

Data Would Increase Understanding of Trends in Power Plant Water Use, GAO-10-23, Oct.16, 2009.

[18] Past forecasts of national water use have proven highly inaccurate. The draft report cited below includes a graph showing a wide variation of previously projected freshwater withdrawals. It further illustrates that all but one overestimated actual water withdrawals: U.S. Army Corps of Engineers, Institute of Water Resources, *Plan of Study: National Water Demand and Availability Assessment*, DRAFT, Alexandria, VA, Feb. 1995, p. 7. Some recent reports, like Forest Service 2010, indicate a role for scenario-based forecasts.

[19] W. B. Solley et al., *Estimated Use of Water in the United States in 1995* (USGS Circular 1200, 1998), http://water.usgs.gov/watuse/pdf1995/html/. Hereafter referred to as USGS 1998.

[20] The Forest Service 2010 water assessment is anticipated to assess the vulnerability of the coterminus United States to water supply shortages over the next 50 years, including alternative scenarios for climate change. It is not breaking out energy's water use or evaluating the sensitivity of the assessments to various policy and technology choices. For more information on this assessment, see http://www.fs.fed.us/rm/human-dimensions/staff /2010_rpa_assessment.shtml.

[21] Elcock 2010; D. Elcock, *Baseline and Projected Water Demand Data for Energy and Competing Water Use Sectors*, Argonne National Laboratory, Environmental Science Division, November 2008.

[22] U.S. Dept of the Interior, "Secretary Salazar Launches New Regional Climate Science Center and Water Census at Meeting of Colorado River Basin Water Leaders," press release, Oct. 20, 2010, http://www.doi.gov/news/pressreleases/ Secretary-Salazar-Launches-New-Regional-Climate-Science-Center-and-Water-Census-at-Meeting-of-Colorado-RiverBasin-Water-Leaders.cfm.

[23] For more information see, the Western Governors' Association, *Regional Transmission Expansion Project*, http://www.westgov.org/index.php?option=com_content&view= article&id=311&Itemid=81.

[24] USDA Forest Service, *Past and Future Freshwater Use in the United States: A Technical Document Supporting the 2000 USDA Forest Service RPA Assessment*, RMRS-GTR-39, Sept. 1999, p. 44.

[25] The extent to which increased water use for bioenergy results in a net increase in national water consumption remains the subject of debate.

[26] Calculated using data tables for Yi-Wen Chiu, Brian Walseth, and Sangwon Suh, "Water Embodied in Bioethanol in the United States," *Environmental Science & Technology,* vol. 43, no. 8 (2009), pp. 2688-2692, hereafter referred to as Chiu 2009; and Kansas Department of Agriculture, Division of Water Resources, *Water Use in Kansas*, http://www.ksda.gov /appropriation/content/116. Insufficient data are available to analyze more local effects on water use (e.g., county level water use). Chiu 2010 includes water used for irrigation and for converting the feedstock into fuel; it does not include precipitation, soil moisture, and other water indirectly consumed by the feedstock.

[27] Forest Service 2010, pp. 15-16, and 24.

[28] Ibid.

[29] DOE, *20% Wind Energy by 2030: Increasing Wind Energy's Contribution to U.S. Electricity Supply*, July 2008, p. 185, http://www1.eere.energy.gov/windandhydro/pdfs/41869.pdf, hereafter referred to as DOE 2008. In the report, the Midwest/Great Plains region included Illinois, Indiana, Iowa, Kansas, Michigan, Minnesota, Missouri, Nebraska, North Dakota, Ohio, Oklahoma, South Dakota, Texas, and Wisconsin.

[30] EIA, Table 1.1, Net Generation by Energy Source, http://www.eia.gov/cneaf/electricity /epm/table1_1.html.

[31] For more on carbon capture, see CRS Report R41325, *Carbon Capture: A Technology Assessment*, by Peter Folger

[32] NETL 2009, p. 64.

[33] USGS 1998, p. 54.

[34] NREL 2003.

[35] The NREL 2003 report included data on the average water consumption per unit of hydroelectric generation by state; the water consumption intensity for the hydroelectric produced at these facilities, according to the study, can vary from a low of 2,000 gallons/megawatt-hour (MWh)(Nebraska) to a high of 154,000 gallons/MWh (Kentucky). The report calculated the weighted average rate of water consumption per MWh at 18,000 gallons, with the variation in the rates across states being significant; the report calculated the national weighted average for thermoelectric generation at 470 gal/MWh.

[36] DOE, Office of Fossil Energy, *Modern Shale Gas Development in the United States: A Primer*, April 2009, p. ES-2, http://www.netl.doe.gov/technologies/oil-gas/publications/EPreports /Shale_ Gas_Primer_2009.pdf.

[37] Ibid, p. 64. An Army Corps of Engineers draft report put the lower end of water use at 0.8 million gallons, with the lower end dominated by the shorter vertical wells. U.S. Army Corps of Engineers, Omaha District, *Garrison/Lake Sakakawea Project, North Dakota, Draft Surplus Water Report*, Omaha, NE, Dec. 2010, p 3-2.

[38] M. Mantell, "Abundant, Affordable, and Surprisingly Water Efficient," presented at 2009 GWCP Water/Energy Sustainability Symposium in Salt Lake City, September 13-16, 2009, available at http://www.energyindepth.org/wpcontent/uploads/2009/03 /MMantell_GWPC_ Water_Energy_Paper_Final.pdf; and EIA, *Average Heat Rates by Prime Mover and Energy Source*, January 2010, http://www.eia.doe.gov/cneaf/electricity/epa/epat5p4.html.

[39] DOE, *Energy Demands on Water Resources: Report to Congress on the Interdependency of Energy and Water*, Dec. 2006, http://www.sandia.gov/energy-water/docs/121-RptTo Congress-EWwEIAcomments-FINAL.pdf, hereafter referred to as DOE 2006.

[40] L. P. Galusky, *An Update and Prognosis on the Use of Fresh Water Resources in the Development of Fort Worth Basin Barnett Shale Natural Gas Reserves*, Texerra Engineering, prepared for Barnett Shale Education Council and Barnett Shale Water Conservation and Management Committee, Midland, TX, Nov. 4, 2009.

[41] Ibid.

[42] C.J. Vavra, "Desalination of Oil Brine," presentation, http://www.pe.tamu.edu/gpri-new/home/BrineDesal/ MembraneWkshpAug06/Burnett8-06.pdf

[43] CRS Report R40631, *Water Issues of Concentrating Solar Power (CSP) Electricity in the U.S. Southwest*, by Nicole T. Carter and Richard J. Campbell.

[44] Ibid.

[45] Dish engine CSP facilities employ engines instead of steam turbines. Mirrors concentrate sunlight to produce heat in a gas chamber connected to a piston and drive shaft. The drive shaft powers an electricity generator. Because of the high operating temperature and high efficiency of the motor, air cooling can be used with little compromise of overall electricity generation efficiency. This significantly reduces the water used to generate electricity compared to other CSP technologies. One dish engine facility with a capacity of 2 MW was put into operation in 2010 in Arizona; dish engine facilities of 750 MW and 850 MW are under development in southern California.

[46] This section on biofuels was written by Megan Stubbs, Analyst in Agricultural Conservation and Natural Resources Policy, Resources, Science and Industry Division (7-8707).

[47] For more information and a map of the Ogallala aquifer, see V. L. McGuire, *Changes in Water Levels and Storage in the High Plains Aquifer, Predevelopment to 2005*, USGS, USGS Face Sheet 2007-3029, 2007, http://pubs.usgs.gov/fs/ 2007/3029/.

[48] Calculated using supplementary data tables from Chiu 2009, and City of Fort Collins, *Water Supply and Demand*, Fort Collins, CO, http://www.fcgov.com/water/supply.php.

[49] S. E. Powers, R. Domingues-Faus, and P. J. J. Alvarez, "The Water Footprint of Biofuel Production in the USA," *Biofuels*, vol. 1, no. 2 (2010), pp. 255-260.

[50] GAO, *Energy Water Nexus: Many Uncertainties Remain about National and Regional Effects of Increased Biofuel Production on Water Resources* (GAO 10-116, Nov. 2009), p. 10. Agricultural runoff with excess nutrients from fertilizers used in the Mississippi River basin contributes significantly to hypoxic "dead zones" in the Gulf of Mexico. For more information on dead zones, see CRS Report 98-869, *Marine Dead Zones: Understanding the Problem*, by Eugene H. Buck.

[51] C.W. King,, M. E. Webber, and I.J. Duncan, "The Water Needs for LDV Transportation in the United States," *Energy Policy*, vol. 38 (2010), p. 1161, hereafter referred to as King 2010.

[52] In particular, the renewable fuels standard (RFS), a major federal incentive established in the Energy Independence and Security Act of 2007 (EISA, P.L. 110-140), establishes a goal of 36 billion gallons of biofuel production by 2022, including 16 billion gallons of cellulosic biofuels. For more information on the RFS, see CRS Report R40155, *Renewable Fuel Standard (RFS): Overview and Issues*. Also, the Food, Conservation, and Energy Act of 2008 (the 2008 farm bill, P.L. 110-246) supports the growth of the cellulosic industry through research programs, grants, loans, and tax credits. For more information on energy provisions in the 2008 farm bill, see CRS Report RL34130, *Renewable Energy Programs in the 2008 Farm Bill*, by Megan Stubbs.

[53] K.R. Fingerman et al., "Accounting for the Water Impacts of Ethanol Production," *Environmental Resources. Letters*, vol. 5 (2010), http://iopscience.iop.org/1748-9326/5/1/014020/pdf/1748-9326_5_1_014020.pdf. Others also have raised concerns that large-scale deployment of open pond algae systems in arid areas may have a freshwater footprint if impaired waters are not used (C. Harto, R. Meyers, and E. Williams, "Life Cycle Water Use of Low-Carbon Transport Fuels," *Energy Policy*, vol. 38 (2010), pp. 4933-4944).

[54] For more information, see CRS Report R41460, *Cellulosic Ethanol: Feedstocks, Conversion Technologies, Economics, and Policy Options*, by Randy Schnepf.

[55] For example, dry cooling is not equally effective for power plant cooling in locations with temperatures above 100°F, or substituting municipal wastewater may not be an option for remote facilities.

[56] J. Leshy, "Notes on a Progressive National Water Policy," *Harvard Law & Policy Review*, vol. 3, no. 1 (Winter 2009), pp. 133-159.

[57] D. H. Getches, "The Metamorphosis of Western Water Policy: Have Federal Laws and Local Decisions Eclipsed the States' Role?," *Stanford Environmental Law Journal*, vol. 20 (2001), pp. 3-72.

[58] For more on rare minerals in the energy sector, CRS Report R41347, *Rare Earth Elements: The Global Supply Chain*, by Marc Humphries.

[59] Elcock 2010.

[60] Elcock 2010 considered that only energy crops would generate an increase in water consumption. The study's projections for water consumption were based on projections for different biofuels, including corn-based ethanol, cellulosic ethanol, and biodiesel.

[61] King 2010, p. 1162.

[62] NETL, *Estimating Freshwater Needs to Meet Future Thermoelectric Generation Requirements*, 2007.

[63] Elcock 2010, p. 451. For more, see CRS Report RL34738, *Cellulosic Biofuels: Analysis of Policy Issues for Congress*, by Kelsi Bracmort et al.

[64] Argonne National Laboratory released for comment the following report: ANL, *DRAFT Water Use in the Development and Operation of Geothermal Power Plants*, September 2010, http://www1.eere.energy.gov/geothermal/ pdfs/geothermal_water_use_draft.pdf, hereafter referred to as Argonne 2010. The draft links its geothermal water use findings with water intensities of other electricity generation. The report cited references from the early 1990s for fuel processing, and those references cited materials from the late 1970s to the early 1990s. Other articles also use data from these sources (e.g., V. Fthenakis and H.C. Kim, "Life-cycle Uses of Water in U.S. Electricity Generation," *Renewable and Sustainable Energy Reviews*, vol. 14 (2010), pp. 2039-2048). Continued reliance on older sources demonstrates a knowledge gap on current water use in fuel acquisition and preparation. Changes in the U.S. fossil fuel practices (e.g., natural gas extraction through fracturing) in recent decades may have significantly altered water use for fuel.

[65] Chemicals added to the water at a thermoelectric facility to extend equipment life and improve operational efficiency (e.g., demineralize regenerants and rinses to prevent biological growth) can degrade water quality when discharged into the environment. To comply with water quality regulations promulgated under the Clean Water Act, thermoelectric facilities constructed after January 17, 2002, generally use evaporative cooling. EPA has begun revising these rules, which were issued in 1982. Separately under the Clean Water Act, EPA regulates cooling water intake structures, which protect fish from entrainment at the intake point. EPA is developing a rule to address water intake structures for some existing thermoelectric facilities, which is expected to be proposed in 2011; the rule may affect decisions about power plan retirements and cooling technology.

[66] DOE, *Concentrating Solar Power Commercial Application Study: Reducing Water Consumption of Concentrating Solar Power Electricity Generation*, no date,: http://www1.eere.energy.gov/solar/pdfs/csp_water_study.pdf.

[67] The Hewlett Foundation and The Energy Foundation, *The Last Straw: Water Use by Power Plants in the Arid West*, April 2003, pg. 12, http://www.catf.us/publications /reports/The_Last_Straw.pdf. Another source estimates that dry cooling and hybrid wet-dry cooling have capital costs 1.1 to 5 times higher than wet cooling (California Energy Commission, *Comparison of Alternate Cooling Technologies for California Power Plants: Economic, Environmental and Other Tradeoffs*, Consultant Report 500-02-079F, Sacramento, CA, Feb. 2002, p. 1-9, http://www.energy.ca.gov/ reports/2002-07-09_500-02-079F.PDF). This same source provides an example of the cooling system capital costs for a 500 MW combined-cycle power plant — \$3.6 million for a wet cooling tower versus \$25.5 million for a dry cooling system (ibid., p. 5-16 and p. 5-39).

[68] Letter from Ric O'Connell, Renewable Energy Consultant, Black & Veatch, to Environmental Working Group, Renewable Energy Transmission Initiative, June 25, 2008.

[69] The loss in generation efficiency is from insufficient cooling of the turbine exhaust steam, increasing steam turbine back pressure. While efficiency losses also are typical of wet-cooled systems when inlet water temperatures exceed design temperatures, the lower cooling capacity of air versus water makes dry cooling more sensitive to temperature changes and efficiency losses than wet cooling. A power plant with dry cooling can experience a 1% loss in efficiency for each 1°F increase of the condenser, which is limited

by ambient temperatures (C. Kutscher, A. Buys, and C. Gladden, *Hybrid Wet/Dry Cooling for Power Plants*, NREL, presented at Parabolic Trough Technology Workshop, Incline Village, NV, Feb. 14, 2006, http://www.nrel.gov/docs/fy06osti/40026.pdf.). Electricity output is also decreased by the additional electricity requirement of running the fans and pumps of the air cooling system.

[70] While there is interest in using alternative water sources such as coastal waters for cooling, there is concern about harm to marine ecosystems.

[71] For example, the Palo Verde Nuclear Generating Station near Phoenix, AZ, uses wastewater effluent for cooling.

[72] D. Lawrence et al., *Power Plant Engineering – Design and Construction*, (Black & Veatch, Chapman & Hall: New York, 1996).

[73] J. A. Veil., J. M. Kupar, and M. G. Puder, Argonne National Laboratory, *Use of Mine Pool Water for Power Plant Cooling*, September 2003, http://www.ipd.anl.gov /anlpubs /2006/11/57830.pdf.

[74] Whether hydroelectric power generation is a renewable is a subject of debate largely because of the environmental impacts of dams and their reservoirs. For more information on small-scale and low-head hydropower, see CRS Report R41089, *Small Hydro and Low-Head Hydro Power Technologies and Prospects*, by Richard J. Campbell.

[75] EPRI, Assessment of Waterpower Potential and Development Needs (Palo Alto, CA: 2007), http://mydocs.epri.com/ docs/public/000000000001014762.pdf.

[76] Idaho National Laboratory, *Feasibility Assessment of the Water Energy Resources of the United States for New Low Power and Small Hydro Classes of Hydroelectric Plants*, (DOE, Washington, DC: 2006), http://hydropower.inel.gov/ resourceassessment/pdfs /main_ report_appendix_a_final.pdf. This study estimated preliminarily feasible opportunities for increasing U.S. hydroelectric generation with a total potential of 30 MW.

[77] DOE, *National Electric Transmission Congestion Study*, December 2009, http://congestion 09.anl.gov/documents/ docs/Congestion_Study_2009.pdf, presents areas where renewable energy development using existing technologies is constrained by transmission, and other areas where development would be constrained if technologies mature (e.g., offshore wind). The areas renewable energy transmission constrained areas are shown in the report on page 23.

[78] Intermittent electricity sources can be used to meet electricity demand for activities that can be ramped up or down (e.g., hydrogen generation via electrolysis, though this technology is still developing commercially) and to reduce generation from other sources when they are available.

[79] DOE, How an Enhanced Geothermal System Works, Washington, DC, March 31, 2006, http://www1.eere.energy.gov/geothermal/printable_versions/egs_animation.html

[80] USGS, *Substantial Power Generation from Domestic Geothermal Resources*, Reston, VA, Sept. 29, 2008, http://www.usgs.gov/newsroom/article.asp?ID=2027&from=rss_home, pg. 5.

[81] For examples of a geothermal potential map, see the following websites: http://www.nrel.gov /gis/images/geothermal_resource2009-final.jpg, and http://www.azgs.state.az.us/images /geothermal_ 6b.jpg.

[82] B.D. Green and G. Nix, *Geothermal - Energy Under Our Feet*, NREL, NREL/TP-840-40665, Nov. 2008.

INDEX

D

E

I

J

L

M

N

O

S

T

U